CHINESE TEA CULTURE

中华茶文化

中国茶叶博物馆 编著

包 静 主编

文物出版社

图书在版编目（CIP）数据

南方嘉木：中华茶文化 / 中国茶叶博物馆编著；包静主编． -- 北京：文物出版社，2023.11
ISBN 978-7-5010-8141-7

Ⅰ. ①南… Ⅱ. ①中… ②包… Ⅲ. ①茶文化—中国 Ⅳ. ① TS971.21

中国国家版本馆CIP数据核字(2023)第139730号

审图号：GS京(2023)1670号

南方嘉木：中华茶文化

编　　著	中国茶叶博物馆
主　　编	包　静
责任编辑	谷　雨
责任印制	王　芳
责任校对	陈　婧
出版发行	文物出版社
社　　址	北京市东城区东直门内北小街2号楼
邮　　编	100007
网　　址	http://www.wenwu.com
经　　销	新华书店
制版印刷	天津图文方嘉印刷有限公司
开　　本	1270mm×965mm　1/16
印　　张	24.25
版　　次	2023年11月第1版
印　　次	2023年11月第1次印刷
书　　号	ISBN 978-7-5010-8141-7
定　　价	528.00元

本书版权独家所有，非经授权，不得复制翻印

编委会

主　编：包　静

策　展（按姓氏笔画排序，不分先后）：

乐素娜　李竹雨　李　靓　朱　阳　朱慧颖　张　佳

汪星燚　郭丹英　姚晓燕　晏　昕

展览协力（按姓氏笔画排序，不分先后）：

王　宏　王　慧　王慧英　李　昕　朱珠珍　张雨梦

张宵悦　张嵩君　余玉庚　周　彬　金鑫英　胡　忠

赵燕燕　崔　雅　黄　超　蔡嘉嘉

序一

FOREWORD I

茶，源自中国，盛行世界，既是全球同享的健康饮品，也是承载历史和文化的"中国名片"。尤其是随着"中国传统制茶技艺及其相关习俗"列入联合国教科文组织人类非物质文化遗产代表作名录，这不仅向世界展现了我国茶文化的独特魅力，也在世界民族之林彰显了中华民族的文化自信。

Tea, originating in China, now prevails all over the globe. It is a healthy drink shared by the whole world, and even a totem of the profound Chinese history and culture. In particular, with the inclusion of "Traditional tea processing techniques and associated social practices in China" in UNESCO's Representative List of the Intangible Cultural Heritage of Humanity, it has not only shown the unique charm of the Chinese tea culture to the world, but also demonstrated the cultural confidence of the Chinese nation in the family of nations.

作为茶叶大国，我国的茶园面积和茶叶产量虽居世界第一，但出口量位列第二。如何借着非遗的东风，进一步推进中国茶文化在世界的广泛传播，如何做好茶文化、茶产业、茶科技三者统筹发展这篇大文章，是茶界需要重点思考的问题。如果把茶产业比作飞机，茶文化和茶科技则是飞机的两翼，有力地促进和保障茶产业的腾飞。近年来，随着科技赋能产业，中国茶的规模和效益有了稳步增长，茶类结构更趋均衡，产品种类丰富多元，茶业也与旅游、健康、文化、互联网等产业不断融合，新业态不断涌现。茶产业已然成为富民产业、生态产业、健康产业和文化产业。

China, a major producer and consumer of tea, boasts the largest tea plantation area and yield in the world, but its export comes in second. How can we take advantage of the intangible cultural heritage to further spread the Chinese tea culture around the world? How do we balance the development of tea culture, tea industry, and tea technology? These are the challenges. If the tea industry is an airplane, the tea culture and tea technology are its two wings to effectively facilitate its take-off. In recent years, with the technology-enabled development of the industry, Chinese tea has been characterized by steadily increased scale and benefits, more balanced tea structure, and rich and diversified product species. The tea industry continues to embrace tourism, health, culture, Internet and other sectors, and new business forms keep emerging. Tea has, nowadays, been a cash crop and a green, health-oriented and cultural industry.

茶产业的繁荣发展不仅需要科技赋能，更离不开茶文化的助推。2022年12月，中国茶叶博物馆完成全面提升改造工程，以全新的面貌向公众开放，在继承原展览优秀经验的基础上锐意创新，改陈后的展览内容更加完善，展示手法更加多样，文物展品更加多元，观众体验更加生动，全方位、多维度、深层次地讲述中国茶的前世今生。同时，为了让公众参观的同时"把博物馆带回家"，中国茶叶博物馆推出《南方嘉木——中华茶文化》一书，全面翔实地介绍展览的内容和文物，弥补公众现场参观展厅时意犹未尽的小遗憾。全书由"草木菁华""茶史寻踪""技艺礼俗""茶传五洲"四部分组成。"草木菁华"从茶树起源、认识茶树到茶叶分类、加工、品饮、储存，再到茶的综合利用、茶与人体健康的关系，重点讲述茶叶的今生故事；"茶史寻踪"讲述茶的前世，按历史年代的脉络阐述历代茶文化的特点；"技艺礼俗"则把视角放在非遗，以图文并茂的方式讲述列入联合国教科文组织人类非物质文化遗产代表作名录的"中国传统制茶技艺及其相关习俗"；"茶传五洲"则讲述了中国茶对世界的贡献和影响力。学术含量极高，体现了国字号专题博物馆的研究水平，不失为历年茶文化图书之精品。

The prosperity of tea industry cannot do without technology, and even the facilitation of tea culture. In December 2022, China National Tea Museum was face-lifted and reopened to the public with a new look. In addition to the original collections, the Museum offers more showcases, display techniques and cultural exhibits, and more interaction with visitors, to tell the entire history of Chinese tea from various angles. Meantime, to enable the public to "bring the museum home", the Museum also launched this illustrated book – *Fine Plants in the South: Chinese Tea Culture*. It gives a comprehensive account of the exhibits to make up the remaining blind spots of visitors when the tea journey is over. This book consists of four chapters: "Tea as a Marvellous Plant", "Tracing the History of Tea", "Tea Processing Techniques and Drinking Practices", "Worldwide Spread of Tea". The first one elaborates the present, ranging from the origin and basic knowledge of tea to the classification, processing, drinking, storage, integrated utilization of tea, and relationship between tea and human health. The second chapter refers to the past, chronologically expounding the characteristics of tea culture in previous dynasties. The third spotlights inheritance, elucidating the "Traditional tea processing techniques and associated social practices in China". The last chapter wraps up the contribution of the Chinese tea to the world and its global influence. The book loaded with tea knowledge and academic findings deserves a thumb-up.

茶叶作为中华文明的载体，不仅在历史上扮演重要的角色，茶更是穿越历史、跨越国界，受到世界人民的喜爱，成为中华文明与世界其他文明交流互鉴的重要媒介，成为人类文明共同的财富。中国茶叶博物馆传播茶文化之事业任重道远。

Tea, a symbol of the Chinese civilization, has played a vital role in human history. It has won a tremendous popularity all over the world. Tea inspires mutual understanding between China and the rest civilizations and has become a shared wealth for the entire humanity. China National Tea Museum is on the right track.

是为序！
Thank you.

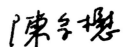

中国工程院院士
Chen Zongmao, member of Chinese Academy of Engineering

序二

FOREWORD II

茶是一种古老而神奇的植物，它的发现和利用无疑是中华民族对人类文明的一大贡献。当我们翻开中华文明上下五千年的历史画卷，可以发现几乎在每一历史阶段都氤氲着茶香。当我们俯瞰中华大地，无论是色、香、味、形各异的茶品，还是气象万千、多姿多彩的茶俗，都生动见证了中华民族永无止境的勤劳和创造，体现了中国人民对美好生活的向往和对身心健康的追求。

Tea is an ancient and mystic plant. Its discovery and proper use are undoubtedly a great contribution of the Chinese people to human civilization. The 5,000-year Chinese history is associated with this magical plant almost at all stages. The myriad of teas in color, aroma, taste and shape, as well as the miscellany of tea customs, have testified we Chinese nation's diligence and wisdom, and embodied our yearning for a better life and pursuit of physical and mental health.

茶是中华传统文化的瑰宝。从种茶、制茶到泡茶、品茶，博大精深的茶技艺、茶习俗凝聚了中华民族的世代匠心，成为延绵数千年生生不息的一种集体记忆。2022年11月29日，联合国教科文组织将"中国传统制茶技艺及其相关习俗"列入人类非物质文化遗产代表作名录，让神奇的"东方树叶"进一步成为弘扬中华优秀传统文化的耀眼明珠。

Tea is a gem of the Chinese culture. Tea-making techniques, like planting, processing and brewing, have remained active as a national memory for thousands of years. On November 29, 2022, "Traditional tea processing techniques and associated social practices in China" was officially added to UNESCO's Representative List of the Intangible Cultural Heritage of Humanity, making it a token of the profound traditional Chinese culture.

从古代丝绸之路、茶马古道，到今天的丝绸之路经济带、21世纪"海上丝绸之路"，茶穿越历史、跨越国界，已经成为中国人民与世界人民相知相交、中华文明与世界其他文明交流互鉴的重要媒介。党的二十大报告指出，要"提炼展示中华文明的精神标识和文化精髓，加快构建中国话语和中国叙事体系，讲好中国故事、传播好中国声音"。讲好数千年的中国茶故事，正是讲好中国故事的精品题材，是传播好中国声音的绝佳频道。

Like that in the past we had the Silk Road and the Old Tea-Horse Road , today we're calling for the building of the Silk Road Economic Belt and the 21st Century Maritime Silk Road. Tea is an envoy to the rest of the world, inspiring more exchanges and mutual understanding between China and other civilizations. As is stated in the report to the 20th National Congress of the Communist Party of China that, "We will collect and refine the defining symbols and best elements of Chinese culture and showcase them to the world. We will accelerate the development of China's discourse and narrative systems, better tell China's stories, make China's voice heard". Tea must be a first-choice for telling such a "good story".

中国茶叶博物馆是我国唯一以茶和茶文化为主题的国家一级博物馆，是展示与传播中华茶文化的重要窗口。由中国茶叶博物馆编著的这部书籍，以馆藏文物和研究成果为依托，从"草木菁华""茶史寻踪""技艺礼俗""茶传五洲"四个篇章对中华茶文化进行全面解读，准确严谨，图文并茂，对普及茶知识、推广茶文化起到了积极的推动作用，相信本书会成为广大读者了解中华茶文化的重要参考资料，同时也会为弘扬中华文化做出积极贡献。

China National Tea Museum is our country's first-class level museum dedicated to tea and tea culture. It should be a great channel of facilitating tea's worldwide popularity – so is the book it prepared. Being the fruits of the Museum's years of collection and academic research, this book will lead the readers to a commanding view of the Chinese tea culture in four chapters, i.e, "Tea as a Marvellous Plant", "Tracing the History of Tea", "Tea Processing Techniques and Drinking Practices", "Worldwide Spread of Tea". Accurate and illustrated, the book must be a new drive for spreading tea knowledge and customs. I believe it would be a significant reference in learning tea's cultural profoundness, which must be a great part of the further promotion of the Chinese culture to the globe.

乐之为序！
Honored to give my plaudits.

刘仲华

中国工程院院士
Liu Zhonghua, member of Chinese
Academy of Engineering

序三

FOREWORD Ⅲ

中国茶文化是中华文明的重要标识之一。

The tea culture is one of the significant symbols of Chinese civilization.

茶促进了世界文明交流，在人类社会可持续发展中发挥着重要作用。

Tea facilitated the exchange of world civilizations and has played an important role in the sustainable development of humanity.

博物馆是历史的保存者和记录者，更是保护和传承人类文明的重要殿堂。中国是茶文化的发源地，也是茶文化最早的传播者。自古以来，茶在我国人民生活中是不可或缺的一部分。

Museums are the preservers and recorders of history, and are also shrines for the continuity of human civilizations. China is the birthplace of tea culture and the earliest country to spread it over the globe. Tea has been, since very early on, an indispensable part of people's lives in China.

中国茶叶博物馆作为以茶和茶文化为主题的国家一级博物馆，始终以讲好茶故事、传播茶文化为己任。当下我们恰逢中国茶业发展盛世，社会公众对茶文化抱有极大的学习热情，以更丰富的内容、更人性化的展陈技术来讲述中国茶故事成为当务之急。2022年8月，我馆正式启动了为期四个月的全面提升改造，2022年12月竣工试开放。全新推出的基本陈列全方位展示了茶文化的丰富内涵。其中"南方嘉木——中华茶文化展"以"草木菁华""茶史寻踪""技艺礼俗"以及"茶传五洲"四大板块内容展开，叙述中国茶文化的发展历史及中国茶对世界的影响，既连接了过去，又沟通了世界，对传承中国茶文化，促进世界茶文化交流互鉴有着重要的意义。

As the country's first-class museum dedicated to tea and tea culture, China National Tea Museum takes the responsibility of tea knowledge education and legacy inheritance. As the tea economy is in the prime now, the public are eager to learn about tea and the culture behind. It's been a pressing mission to educate them with instructive displays. The upgrading of museum facilities was launched in Aug. 2022, which has costed 4 months and was completed in December. The new collections offer an all-encompassing perspective of tea culture, in which the *Fine Plants in the South: Chinese Tea Culture* is divided into four sections, namely, "Tea as a Marvellous Plant", "Tracing the History of Tea", "Tea Processing Techniques and Drinking Practices", "Worldwide Spread of Tea". It tells the history of Chinese tea culture and its impact on the rest of the world, and shows how the magic beverage connected the past and eventually reached the globe. It's been of great value for the mutual learning of the world's tea legacies.

未来的我们将继续发挥博物馆自身特色优势，用茶文物讲好中国故事，为公众提供更加丰富、独特、有趣的精神文化食粮，充分展现中国茶历史文化的精彩，让中华茶文化散发出永久魅力和时代风采。

In the future, we will continue to give full play to our strengths and work hard on providing visitors with an insight into the culture. We will showcase with all our gifts about the splendor of China's tea history and culture and let them shine for good.

啜苦咽甘，茶也。如今的中国茶叶博物馆灿然现于世人，改陈之辛苦已然化为满心甘甜。感谢为之辛勤付出的领导、专家及团队成员们，是以集结策展之成果，以备后鉴。

Tea tastes bitter, yet then sweet. It's like our efforts that went all through hardship before today's fruits. I thank everyone in support of the museum, and I wish the exhibition a tryout for more event to come.

是为序。
Thanks.

包 静

中国茶叶博物馆馆长
Bao Jing, Director of China National Tea Museum

前言 / 001
PREFACE

第一篇
草木菁华

PART I
TEA AS A MARVELLOUS PLANT

第二篇
茶史寻踪

PART II
TRACING THE HISTORY OF TEA

第一章	茶树起源 *Origin of Tea Trees*	009
第二章	认识茶树 *Introduction to Tea Trees*	012
第三章	茶叶加工 *Tea Processing*	021
第四章	茶叶品类 *Tea Category*	025
第五章	茶叶品饮与储存 *Tea Drinking and Storage*	044
第六章	茶与健康 *Tea and Health*	053

第一章	茶生南国——饮茶的起源 *Tea Grows in Southern China Tracing the Origin of Tea Drinking*	063
第二章	茶风初扬——先秦至魏晋南北朝茶文化 *Formative Period of Tea Culture Tea Culture from Pre-Qin to Wei, Jin, Southern and Northern Dynasties*	070
第三章	茶事盛景——唐代茶文化 *Flourishing Tea Culture Tea Culture in the Tang Dynasty*	082
第四章	茶为清尚——宋元茶文化 *Drinking Tea, an Elegant Fashion Tea Culture in Song and Yuan Dynasties*	114
第五章	茶韵隽永——明代茶文化 *Profound Tea Culture Tea Culture of the Ming Dynasty*	150
第六章	茶意不歇——清代茶文化 *Endless Charm of Tea Tea Culture of the Qing Dynasty*	162
第七章	茶业复兴——近现代茶文化 *Revival of Tea Industry Modern and Contemporary Tea Culture*	198

PART III
TEA PROCESSING TECHNIQUES AND DRINKING PRACTICES

第一章　唯有佳茗　不负光阴 —— 213
A Sip of China and the World

第二章　味久而淳　香远益清 —— 215
The Source of Thrills in Smell and Taste

第三章　寻茶问水　咀华啜英 —— 304
Tea Drinking Practices

PART IV
WORLDWIDE SPREAD OF TEA

第一章　芳茶远播 —— 315
Spread of Tea and Tea Culture

第二章　五洲茶话 —— 332
Tea Culture of the Five Continents

第三章　世界茶业 —— 365
Tea Industry in the World

结　语 / 375
EPILOGUE

PART I

TEA AS A MARVELLOUS PLANT

第四章

茶叶品类
Tea Category

025

一 六大基本茶类　Six Categories of Tea …… 026

二 再加工茶　Reprocessed Tea …… 036

第五章

茶叶品饮与储存
Tea Drinking and Storage

044

一 沏茶用水　Water for Brewing Tea …… 044

二 沏茶用具　Utensils for Brewing Tea …… 047

三 茶叶冲泡　Tea Brewing …… 049

四 茶叶储存　Tea Storage …… 051

第六章

茶与健康
Tea and Health

053

前言
PREFACE

中国是茶树的原产地，是最早发现和利用茶叶的国家。几千年来，随着饮茶风习不断深入中国人民的生活，茶文化在中国悠久的民族文化长河中不断丰厚和发展，成为东方传统文化的瑰宝。今天，茶作为一种世界性的饮料，维系着中国和世界各国人民的情感。

China is the origin of tea trees, and it is the first country to discover and use tea. For thousands of years, with the custom of drinking tea going into Chinese people's life, tea culture has been enriched and developed in the long river of national culture in China, and has become a treasure of traditional oriental culture. Today, tea, as a worldwide beverage, sustains the feelings of people in China and other countries.

草木

第一篇

第一章

茶树起源
Origin of Tea Trees
009

第二章

认识茶树
Introduction to Tea Trees
012

一 茶树的形态特征 Morphological Characteristics of Tea Tree 012

二 茶树品种 Variety of Tea Plants 016

三 茶树的一生 Life of Tea Tree 017

四 茶树的生长环境 Growing Environment of Tea Plants 018

五 茶树的地理分布 Geographical Distribution of Tea Trees 020

第三章

茶叶加工
Tea Processing
021

一 茶叶采摘 Tea Plucking 021

二 制茶工艺 Tea Producing Process 023

第一章 茶树起源

Origin of Tea Trees

茶源于中国，在人们发现并利用它之前，它便已存在。那么，茶树究竟起源于何时呢？

Tea originated in China and had existed before it was discovered and used. When did tea trees appear?

茶树起源的时间是一个至今仍未十分明确的问题。从植物系统学推论，山茶属茶亚属植物可能出现在距今七千多万年到六千多万年的中生代末期到新生代初期，茶树的出现比人类早数千万年。

It is not quite clear yet when tea trees originated. It is inferred from plant systematics that the subgenus camellia may appear from the late Mesozoic to the early Cenozoic, more than 60 million to 70 million years ago. Tea trees predate humans by tens of millions of years.

关于茶树的起源地，历来争论较多，目前比较一致的看法是：中国是世界茶树原产地，更具体地说，中国西南地区，主要包括云南的东南部和南部、广西的西北部、贵州的西南部是茶树原产地的中心。据不完全统计，现在中国已有10个省（自治区、直辖市）三百多处发现有野生大茶树。

图1.1 新生代早第三纪四球茶茶籽化石（贵州省农业科学院茶叶研究所藏）
Fossil seeds of four-ball tea from Early Tertiary in Cenozoic, Tea Research Institute of Guizhou Academy of Agricultural Sciences

There has always been a lot of controversy about the origin of tea trees. The widely-accepted view is that China is home to tea trees. More specifically, the southwest of China, including the southeast and south of Yunnan, the northwest of Guangxi and southwest of Guizhou, is the center of the original tea trees. Incomplete statistics show that more than 300 wild tea trees have been found in 10 provinces and cities (districts) in China.

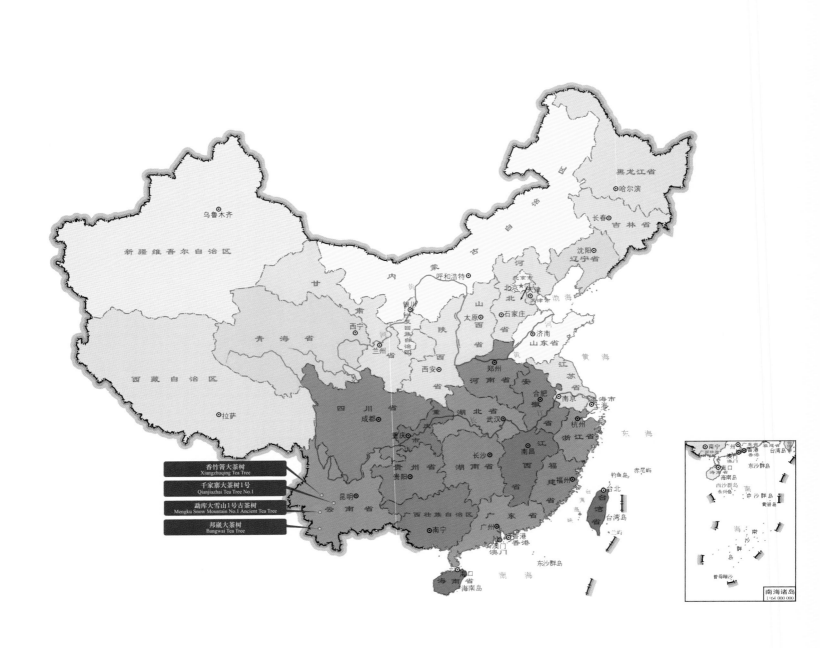

图1.2 中国古茶树分布图

Distribution map of ancient tea trees in China

图1.3 千家寨大茶树1号
Qianjiazhai tea tree No.1

图1.4 勐库大雪山1号古茶树
Mengku Snow Mountain No.1 ancient tea tree

图1.5 邦葳大茶树
Bangwai tea tree

图1.6 香竹箐大茶树
Xiangzhuqing tea tree

第二章 认识茶树

Introduction to Tea Trees

茶是人类采摘茶树上的鲜叶加工制作而成的一种古老且健康的饮品。茶叶的种类繁多，形态各异，产地分布广泛，品质也有诸多不同，但是它们都有一个共同的来源——茶树。所以，要了解各种形态和品质的茶叶，首先应该从认识和了解茶树开始。

茶树是一种多年生、木本、常绿植物。

Tea is an old healthy drink made from fresh leaves plucked from tea trees. There are many kinds of tea, though of different shapes, produced in various places and of distinct qualities, they all have a common source: tea trees. Therefore, to know them well, we should begin with tea trees.

The tea tree is a perennial, woody and evergreen plant.

一

茶树的形态特征
Morphological Characteristics of Tea Tree

在古代，我们的祖先就对茶树有了详细的描述。唐代陆羽在《茶经》中写道："其树如瓜芦，叶如栀子，花如白蔷薇，实如栟榈，蒂如丁香，根如胡桃。"茶树是由根、茎、叶、花、果实、种子等器官有机结合的一个整体，共同完成新陈代谢和生长发育过程。

In ancient times, our ancestors described the tea tree in detail. As described in Lu Yu's *Classic of Tea*, "The general appearance of the tea plant is similar to the evergreen Ilex latifolia Thunb, though its leaves are more like a gardenia's, its white flowers like the rosette's, seeds like the palm tree's, pedicles like the clove's, and roots like the walnut's." Tea tree is an organic whole composed of roots, stems, leaves, flowers, fruits, seeds and other organs, which together complete the process of metabolism, growth and development.

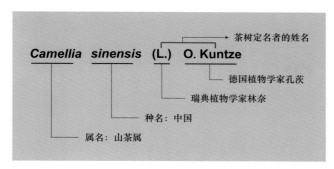

图1.7 茶树学名
The scientific name of tea plant

图1.8 茶树在植物分类学上的地位
Position of tea tree in Plant Taxonomy

图1.9 三种不同形态的茶树
Tea plants in three shapes

● 根 Root

茶树的根由主根、侧根、吸收根、根毛组成。主根和侧根呈红棕色，寿命长，起固定、贮藏和输送作用；吸收根主要吸收水分和无机盐，寿命短；吸收根的表面密生着根毛。

The root system is composed of the main root, lateral roots, absorption roots and root hairs. The former two are reddish brown in color. They have a long lifespan, playing a role in fixation, storage, and transportation. Absorbing roots mainly absorb water and inorganic salts, but they usually have a short lifespan. The surface of the absorbing roots is densely covered with root hairs.

● 茎 Stem

茎是联系茶树根与叶、花、果，输送水、无机盐和有机养料的轴状结构。根据茶树茎的分枝部位不同，茶树可分为乔木、小乔木、灌木三种类型。

Stem is an axial structure connecting roots, leaves, flowers and fruits, and conveying water, inorganic salts, and organic nutrients. According to the difference in stem branching, tea plants can be divided into three types: arbor tea trees, small-arbor tea trees, and shrubs.

● 叶 Leaf

叶是茶树进行光合作用、制造养分的营养器官，也是人们采收利用的对象。

Leaves, the vegetative organs of tea plants for photosynthesis and nutrient production, are also the objects of people's harvesting and utilization.

1. 主脉明显，侧脉呈≥45°角伸展至叶缘三分之二的部位，向上弯曲与上方侧脉相连接。

2. 叶缘有锯齿，锯齿呈鹰嘴状，一般16～32对，随着叶片老化，锯齿上腺细胞脱落，并留有褐色疤痕。

01 / 乔木型 *Arbor tea tree*

此类茶树植株高大，主干明显，分枝部位高。自然生长状态下，树高通常达3~5米，野生茶树高达10米以上。如云南大叶种、海南大叶茶，以及大部分野生大茶树等。

This kind of tea trees is tall, with obvious trunk and high branches. Growing in the natural state, usually the trees will be 3-5 meters high, and wild tea trees will be more than 10 meters, such as Yunnan big leaf species, Hainan big leaf tea, and most wild big tea trees.

树体：主干高大，主干、侧枝、分枝明显。
分布：云南、海南、广西一带。
茶树类型：较原始的茶树类型，多为野生古茶树。

Main characteristics: Tall, with obvious trunk, laterals and branches.
Distribution: Yunnan, Hainan and Guangxi.
Type: Relatively primitive tea tree. Most wild ancient tea trees are in this group.

02 / 小乔木型 *Small-arbor tea tree*

亦称"半乔木型"，此类茶树植株高度中等，分枝部位离地面较近，主干较明显。小乔木型茶树在自然生长状态下，树冠较高大，如政和大白茶、福鼎大白茶等。

It is also called "semi-arbor tea tree". This kind of tea trees of medium height, its branches are close to the ground, and its trunk is obvious. In the natural growth state, small-arbor tea trees have tall crowns, such as Zhenghe White Tea and Fuding White Tea.

树体：主干明显，主干和分枝容易区分，
　　　但分枝距离地面较近。
分布：广东、台湾、福建一带。
茶树类型：较早期进化类型，我国主要栽培品种。

Main characteristics: This type features conspicuous trunks, and distinguishable trunks and branches, yet the latter is close to the ground.
Distribution: Guangdong, Taiwan and Fujian.
Type: The early evolution type, also the main variety cultivated in China.

图1.10 三种不同形态茶树的特征
Characteristics of tea plants in three shapes

1. The main vein is obvious, and the side veins extend to 2/3 of the leaf edge at an angle of ≥ 45°, and bend upward to connect with the upper side veins.

2. The leaf margin has beak-shaped serrations, which are usually 16–32 pairs. As the leaves age, the glandular cells on the serrations fall off, leaving brown scars.

3. 嫩叶背面着生茸毛。

3. The back of the tender leaves is covered with fuzzes.

03 / 灌木型 *Shrub*

此类茶树植株矮小，自然生长状态下，树高通常只达1.5~3.0米。近地面处枝干丛生，分枝稠密，成年后无明显主干，我国栽培的茶树多属此类。

This kind of tea tree is short, and its height is usually only 1.5-3.0 meters when it grows naturally. The branches near the ground are clustered densely, and there is no obvious trunk in adulthood. Tea plants cultivated in China mostly fall into this category.

树体：主干矮小，主干和分枝不明显，分枝较稠密。
分布：长江南北广大茶区。
茶树类型：进化类型，我国主要栽培品种。

Main characteristics: Short trunk; inconspicuous trunk and branches; dense branches.
Distribution: Tea areas in the north and south of the Yangtze River.
Type: Evolutionary type, the main variety cultivated in China.

● 花 Flower

茶树花属两性花，花冠为白色，少数呈淡绿或粉红色，通常由5~9片花瓣组成，分2层排列。花芽于每年的6月中旬形成，10~11月为盛花期。

Tea flowers are bisexual, with white corolla, a few pale green or pink, usually consisting of 5－9 petals in 2 layers. Flower buds come up in mid-June every year, and the flowering period is from October to November.

● 果实与种子 Fruit and Seed

茶果为蒴果，成熟时果壳开裂，种子落地。果皮未成熟时为绿色，成熟后变为棕绿或绿褐色。茶果形状和大小与茶果内种子粒数有关。茶籽一般在11月前后采收。

The fruit is a capsule, and when it is ripe, the shell cracks and the seeds fall to the ground. The peel is green when immature, and turns brown or green-brown when mature. The shape and size of tea fruit are related to the number of seeds in tea fruit. Tea seeds are generally harvested around November.

图1.11 茶树叶片的基本特点
Features of tea plant leaves

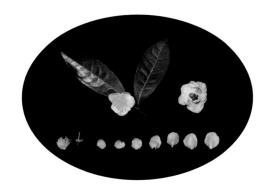

图1.12 茶树花的结构
Structure of tea flowers

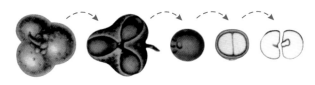

图1.13 茶树果实与种子
Tea fruits and seeds

二

茶树品种
Variety of Tea Plants

中国是世界上茶树种质资源最为丰富的国家，悠久的产茶史为孕育大批优良品种提供了条件。据统计，中国已育成登记的茶树品种有近四百个，且涵盖品类广，满足了生产上的各类需求。

China owns the richest tea germplasm resources in the world, and its long history of tea production provides possibilities for the birth of excellent varieties. Statistics have shown there are nearly 400 registered tea varieties in China, covering a wide range of varieties, meeting all kinds of production needs.

图1.14 紫红叶茶
Tea with purple-red leaves

图1.15 白叶茶
White-leaf tea

图1.16 黄叶茶
Yellow-leaf tea

图1.17 福建省特色茶树新品系基地
Special new tea strain base in Fujian Province

三

茶树的一生
Life of Tea Tree

茶树的生命周期很长，从种子萌芽、生长、开花、结果、衰老、更新直到死亡，要经历数十年到数百年。由于茶树繁殖方式不同，它的一生并不完全一致。例如，有的茶树是由种子萌发生长而成的，有的茶树是由一小段枝条扦插后，生长发育而成的。一株茶树的生长发育过程大致可分为四个时期：幼苗期、幼年期、成年期和衰老期。茶树的一生可达100年以上，而经济生产年限一般只有40～60年。

The life cycle of tea plants is very long, from seed germination, growing, flowering, fruiting, aging, renewal to death, which takes decades to hundreds of years. Different reproduction ways of tea plants cause them to experience different lives. For example, some tea plants grow from seed germination, and some develop after a short branch cutting. The growth and development process of a tea plant can be roughly divided into four stages: seedling stage, juvenile stage, adult stage and senescence stage. The lifetime of tea plants can be as long as over 100 years, while the economic production life is generally only 40 – 60 years.

幼苗期　　　幼年期　　　成年期　　　衰老期

图1.18 茶树的一生
The life of a tea plant

扫码了解
茶树的一生

图1.19 茶树的生长环境
Growing environment of tea plants

四

茶树的生长环境
Growing Environment of Tea Plants

由于茶树原产地的生态环境是热带雨林，这使得它天生就喜欢温暖湿润的气候环境。

Because originally tea trees grew in tropical rain forest, they like warm and humid climate by nature.

● 温度 Temperature

茶树生长最适宜的温度是20～30℃。多数茶树品种日平均气温要稳定在10℃以上，茶芽开始萌动。当气温继续升高到14～16℃时，茶芽逐渐展开成嫩叶。当平均气温低于10℃时，茶芽停止萌发，处于越冬休眠状态。当然，如果气温高于40℃时，茶树也会容易死亡。

● 光照 Light

光照是茶树生存的重要条件，但也不能太强。茶树具有耐荫的特性，在较弱的光照下就能达到较高的光合作用效果。如果光照太强的话，茶叶中苦涩的多酚类化合物含量会增加，影响茶叶的滋味，这也正是中国俗话说"高山云雾出好茶"的道理所在。

Light is an important factor for the survival of tea plants, but it should not be too strong. Shade-tolerant tea plants achieve high photosynthesis in weak light. If the light is too strong, the content of bitter polyphenols in tea will increase, which will affect the taste of tea. This is the reason why the Chinese people say "Good tea comes from cloudy and misty high mountains".

● 水分 Moisture

茶树喜欢湿润，栽培茶树年降雨量最适约1500毫米，生长季节月降雨量要求在100毫米以上，当月降水量少于50毫米时，茶树缺水。空气相对湿度以70%~90%为宜，低于50%对茶树生长发育不利。

Tea plants like a humid environment. The optimum annual rainfall for cultivating tea plants is about 1500 mm. The monthly rainfall in the growing season should be over 100 mm. When the monthly rainfall is less than 50 mm, tea plants will be short of water. The air relative humidity should be 70%–90%, but less than 50% is unfavorable to the growth and development of tea plants.

● 土壤 Soil

茶树是耐酸作物，只有在酸性土壤中才可以正常生长，土壤pH值以4.5~5.5为宜。土壤质地一般以通气、排水性能良好的沙质壤土为好。

Tea plants are acid-resistant. It can grow normally only in acidic soil, and the pH value of the soil is preferably 4.5–5.5. Generally, sandy loam with good ventilation and drainage performance is the best for tea plants.

Tea plants grow best when the temperature is between 20℃ and 30℃. The average daily temperature for most tea varieties should be above 10℃, when tea buds begin to sprout. As the temperature rises continuously to 14–16℃, tea buds gradually expand into tender leaves. When the average temperature is lower than 10℃, tea buds stop sprouting and are in the overwintering dormant state. Tea plants will die easily if the temperature is higher than 40℃.

五

茶树的地理分布
Geographical Distribution of Tea Trees

现今中国茶树的生长区域很广,从北纬18°46′的海南五指山到北纬37°52′的山东烟台,从东经94°15′的西藏林芝到东经121°45′的台湾宜兰都有分布。在中国,通常将产茶区划分为江北、江南、西南、华南四个茶区。

Tea plants grow extensively in China, ranging from Five Finger Mountain in Hainan at 18°46′ N to Yantai in Shandong at 37°52′ N, from Linzhi in Tibet at 94°15′ E to Yilan in Taiwan at 121°45′ E. Chinese tea-producing areas are usually divided into four, namely, North of the Yangtze River, South of the Yangtze river, Southwest and South China districts.

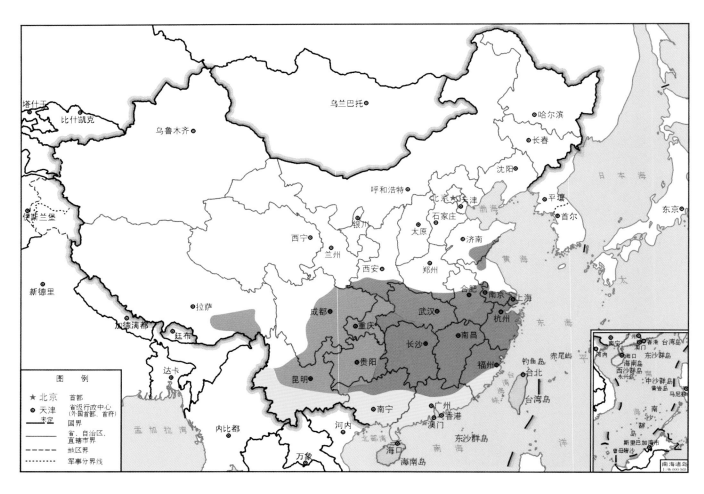

图1.20 四大茶区分布图
Map of four tea-growing areas

- 江北茶区 North-of-Yangtze Region
- 江南茶区 South-of-Yangtze Region
- 西南茶区 Southwest China Region
- 华南茶区 South China Region

第三章 茶叶加工

Tea Processing

茶叶加工是茶叶从茶园到茶杯的重要环节。茶树鲜叶经过各道制茶工序被加工成各种半成品或成品茶，可分为初制、精制、再加工。初制对茶叶品质影响最大，是形成优良品质的基础。茶叶加工是否合理，关系到茶叶的产量、质量和经济效益等。

To make tea a drinkable beverage, processing is the key step. Fresh tea leaves are processed into various semi-finished or finished teas through various tea-making processes, which can be divided into primary preparation, refining and reprocessing. The primary preparation, which establishes the basis for good quality, has the greatest influence on the quality of tea. Whether tea is processed successfully matters greatly in terms of output, quality and economic benefits of tea.

一 茶叶采摘 Tea Plucking

茶叶采摘，即从茶树上采收鲜叶的作业。中国茶区的采摘多有季节性，根据茶芽生长和休止时间可分为春茶、夏茶、秋茶，华南茶区还有冬茶。根据茶叶采摘的方式，主要分为手工采茶和机械采茶。

Tea plucking means harvesting fresh leaves from tea plants. Plucking in China's tea areas is seasonal, which generally happens in spring, summer, autumn and also winter in Southern China. Tea leaves are picked either by hand or machine.

图1.21 手工采茶
Manual tea plucking

图1.22 机械采茶
Mechanical tea plucking

二 制茶工艺
Tea Producing Process

从茶树新梢上采下的芽叶，通过不同的加工方法，可制成不同品质特点的六大茶类：绿茶、黄茶、黑茶、白茶、青茶（乌龙茶）、红茶。

Bud leaves picked from new tea twigs can be processed, using different methods, into six major types of tea: green tea, yellow tea, dark tea, white tea, oolong tea and black tea, each having unique characteristics.

●绿茶加工
Processing of Green Tea

绿茶种类很多，加工方式多样，基本工艺可概括为杀青、揉捻和干燥。

There are many different kinds of green tea. Various processes are required for the manufacture of green tea, but the basic ones are enzyme inactivation, rolling and drying.

●黄茶加工
Processing of Yellow Tea

黄茶加工的基本工艺为杀青、揉捻、闷黄、干燥，形成黄茶品质的关键工序是"闷黄"。

The basic techniques for yellow tea processing are enzyme inactivation, rolling, withering and drying. The third step is crucial for its quality.

●黑茶加工
Processing of Dark Tea

黑茶加工的基本工艺为杀青、揉捻、渥堆、干燥，形成黑茶品质的关键工序是渥堆。

The basic processing techniques for dark tea are enzyme inactivation, rolling, heaping and drying, among which the third step is crucial.

●白茶加工
Processing of White Tea

白茶加工的基本工艺为萎凋、干燥，不炒不揉，加工工艺独树一帜。

White tea is manufactured in simple but special ways, involving neither pan-firing nor rolling. The basic techniques are withering and drying.

●青茶（乌龙茶）加工
Processing of Oolong Tea

青茶（乌龙茶）加工的基本工艺为晒青、晾青、摇青、杀青、揉捻和干燥，其加工结合了绿茶和红茶的制作工艺。

The basic techniques for oolong tea processing are sun-drying, air-drying, rotating, enzyme inactivation, rolling and drying. They are the combination of green tea and black tea manufacturing techniques.

●红茶加工
Processing of Black Tea

红茶加工的基本工艺为萎凋、揉捻、发酵和干燥。发酵是形成红茶品质特征的关键工序。

The basic techniques for black tea are withering, rolling, fermenting and drying, among which fermenting is an essential step.

传统手工制茶和现代机械制茶各有优劣。传统手工制茶与制茶师的技艺有直接关系,能生产出高品质茶,但产量少。现代机械制茶注重标准化、规范化,能大大提高茶叶产量。我们在大力发展机械制茶的同时,也要继承发扬传统制茶技艺,全面推动茶产业高质量发展。

Traditional tea processing techniques and modern mechanical tea production have their own strengths and weaknesses. High-quality tea can be produced manually, though the output is low. Manual processing is influenced by tea producers' skills, while modern mechanical processing focuses on standardization and normalization, which greatly increases tea production. For the overall development of the tea industry, we should both promote vigorously mechanical tea production and inherit and carry forward the traditional tea-making techniques.

图1.23 全自动揉捻机组
Full-automatic rolling machine

扫码了解
现代机械化制茶生产

图1.24 自动智能化生产加工流水线(谢裕大提供)
Smart tea processing assembly line (photo by Xie Yuda)

第四章 茶叶品类

Tea Category

中国产茶历史悠久，产茶区域辽阔。在漫长的生产实践中，中国茶人积累了丰富的茶叶采制经验，历经数千年的发展，产生了花样繁多、品类各异的名茶。制法之精、质量之优、风味之佳，令人叹为观止。从茶树新梢上采下的芽叶，通过不同的加工方法，可制成不同品质特点的六大基本茶类：绿茶、黄茶、黑茶、白茶、青茶（乌龙茶）、红茶。在此基础上可再加工成为花茶、紧压茶、工艺造型茶、萃取茶、末茶等。

Tea production has a long history and there are vast tea-producing areas in China. Over thousands of years, Chinese tea workers have accumulated rich experiences in plucking and manufacturing tea. They have also developed an abundance of famous teas, which are amazing in terms of workmanship, quality, flavor and taste. The bud leaves from the new twigs of tea trees can be made into six categories of tea with different quality characteristics: green tea, yellow tea, dark tea, white tea, oolong tea and black tea. All this can further be processed into scented tea, compressed tea, craft tea, extracted tea, powdered tea, etc.

六大基本茶类
Six Categories of Tea

1. 绿茶 Green Tea

绿茶是中国茶叶产量最多的一类，全国各个产茶区都有绿茶生产。中国绿茶的产量及花色品种居世界之首。绿茶具有绿叶绿汤的品质特征，以加工方式不同，分为炒青绿茶、烘青绿茶、蒸青绿茶、晒青绿茶。

The green tea is produced most in China, and all tea-producing areas in the country yield green tea. Chinese output and products of green tea rank first in the world. Green tea features green leaf and infusion, and can be divided into fried green tea, baked green tea, steamed green tea and sun-dried green tea according to different processing methods.

图1.25　贵州苗岭云雾茶园
Guizhou Miao Ling Yunwu tea garden

图1.26　太平猴魁
Taiping Houkui Tea

图1.27　恩施玉露
Enshi Yulu

图1.28　浙江杭州梅家坞茶园
Zhejiang Hangzhou Meijiawu tea garden

图1.29 浙江温州平阳黄汤茶博园
Zhejiang Wenzhou Pingyang Huangtang tea garden

2. 黄茶 Yellow Tea

黄茶具有黄叶黄汤的品质特征，加工工艺在绿茶工艺基础上多了闷黄工序。黄茶按原料嫩度依次分为黄芽茶、黄小茶和黄大茶。

Leaves and infusion of the yellow tea are both yellow in color. The processing techniques are based on those for the green tea with extra withering process. It can be classified into yellow bud tea, Huangxiaocha and Huangdacha according to the tenderness of raw materials.

图1.30　蒙顶黄芽
Mengding Yellow Bud

图1.31　平阳黄汤
Pingyang Huangtang Tea

图1.32　君山银针杯泡
Junshan Silver Needle in a glass cup

3. 黑茶 Dark Tea

黑茶是中国的特色茶类，主产于湖南、云南、四川、湖北、广西等地，品质特征为叶色油黑或黑褐，汤色褐黄或褐红。黑茶以往主要供边疆地区人们饮用，又称边销茶。进入21世纪后，由于对黑茶功能性成分及其保健作用研究取得重大创新成果，黑茶消费者日益增多。

The dark tea, a special tea in China, is mainly produced in Hunan, Yunnan, Sichuan, Hubei, Guangxi, etc. The leaves are oily black or dark brown, and the infusion brown yellow or brown red. Dark tea, also known as "Bianxiao Tea", used to be consumed by people in border areas. In the 21st century, an increasing number of people have become dark tea drinkers, resulting from the great achievements in the research of its functional components and health care effects.

图1.33 云南茶园
Yunnan tea garden

图1.34 湖南安化芙蓉山茶园
Hunan Anhua Furong Mountain tea garden

图1.35 六堡茶叶底和茶汤
Soaked Liu Bao Tea leaves and tea infusion

4. 白茶 White Tea

白茶是中国的特色茶类,主产于福建。茶叶满披白毫,汤色嫩黄浅淡,滋味清甜甘醇。

The white tea, a special tea in China, is mainly produced in Fujian Province. The tea leaves are covered with white hairs, the infusion is pale yellow, and the taste is sweet.

白毫银针
Baihao Yinzhen (Silver Needles)

白牡丹
Baimudan (White Peony)

寿眉
Shoumei

图1.36 白茶干茶
White dry teas

图1.37 福建福鼎市点头镇生态茶园
Ecological tea garden, Diantou town, Fuding, Fujian province

5. 青茶 Oolong Tea

青茶也叫乌龙茶，是中国的特色茶类，具有绿叶红镶边的特征，汤色橙黄或金黄，主产于福建、广东和台湾。福建乌龙分为闽北和闽南两大产区：闽北乌龙以武夷岩茶为代表，闽南乌龙以安溪铁观音为代表。广东乌龙以凤凰单丛为代表。台湾乌龙以包种茶为代表。

The oolong is a tea unique to China, with green leaves and red edges. Mainly produced in Fujian, Guangdong and Taiwan, its infusion is orange or golden. Fujian oolong is produced both in the north and south of the province, represented respectively by Wuyi Rock Tea and Anxi Tieguanyin. Fenghuang Dancong is the typical oolong tea in Guangdong, just as Baozhong Tea in Taiwan.

图1.38 福建武夷山茶园
Fujian Wuyishan tea garden

图1.39 台湾茶园
Taiwan tea garden

图1.40 大红袍
Dahongpao

图1.41 福建安溪铁观音茶园（李志骐摄影）
Plantations of Fujian Anxi'sTieguanyin(photo by Li Zhiqi)

6. 红茶 Black Tea

红茶具有红叶红汤的品质特征。红茶按制作方法不同，分为小种红茶、工夫红茶和红碎茶三类。小种红茶产于福建，茶叶具有浓郁的松木烟香；工夫红茶以做工精细而得名，按产地不同分祁红、滇红、宁红、闽红等，品质各具特色；红碎茶工艺以切代揉，呈细小颗粒状，滋味浓、强、鲜。

图1.42 滇红
Dianhong Tea

图1.43 正山小种
Lapsang Souchong

图1.44 广东英德茶园
Guangdong Yingde tea garden

The black tea has red leaves and produces red infusion. The black Chinese tea family includes Xiaozhong black tea, Congou black tea and chopped black tea. The Xiaozhong

图1.45 湖北鹤峰木耳山茶园
Hubei Hefeng Mu'er Mountain tea garden

black tea is produced in Fujian. The burning of pine for withering and drying gives it a rich pine smoky flavor. The Congou black tea, so named because of the fine workmanship, has a delicate appearance. It can be classified into Qihong, Dianhong, Ninghong and Minhong according to origin, each having a remarkable quality. The chopped black tea, characterized by thick, strong and fresh taste, looks like fine grains because it is chopped rather than rolled.

二

再加工茶
Reprocessed Tea

1. 花茶 Scented Tea

茶叶成品与鲜花拼和窨制，使茶叶吸收花香。用于窨制花茶的鲜花主要有茉莉花、白兰花、玳玳花、桂花、珠兰花、玫瑰花等。花茶的产区主要分布在广西、福建、四川、云南等地。

The finished tea is blended and scented with fresh flowers, so that the tea can absorb the fragrance of flowers. Flowers used in the scenting mainly include jasmine, hyacinth orinntal, Seville orange flowers, sweet-scented osmanthus, chloranthus, roses, etc. Scented tea is chiefly produced in Guangxi, Fujian, Sichuan, Yunnan and other places.

图1.46　茉莉龙珠
Jasmine Balls

图1.47　桂花龙井
Sweet-osmanthus Longjing

图1.48　茉莉花茶
Jasmine Tea

图1.49　珠兰花茶
Zhulan Scented Tea

2. 紧压茶 Compressed Tea

紧压茶以黑茶、绿茶、红茶、白茶等毛茶为原料，经再加工蒸压成一定形状，主要产于湖南、湖北、四川、云南、广西等省（自治区）。紧压茶质地紧实，久藏不易变质，便于储运，适合边疆牧区人民的需要。如今的紧压茶面对的是全社会的消费者，特别是普洱茶正成为越来越多人钟爱的饮品。

Using dark tea, green tea, black tea, white tea, etc. as raw materials, compressed tea is processed, steamed and pressed into certain shapes. They are mainly produced in Hunan, Hubei, Sichuan, Yunnan, Guangxi and other provinces (or autonomous regions). Compressed tea is very tight, non-perishable, and convenient for storage and transportation, so they are welcomed by people in frontier and pastoral areas. Today's compressed tea is facing the consumers of the whole society, among which Pu'er tea is becoming a favorite drink of more and more people.

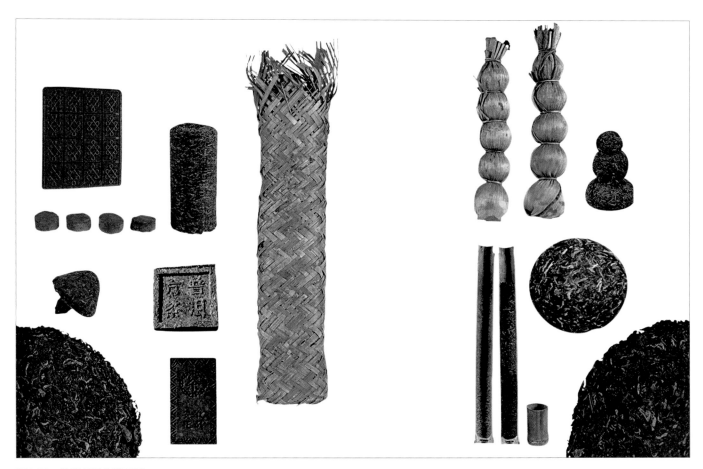

图1.50　造型各异的紧压茶
Compressed teas

图1.51 黄山绿牡丹

Lvmudan (Green Peony) of Huangshan Mountain

3. 工艺造型茶 Craft Tea

工艺造型茶，精选上等茶叶为原料，与脱水鲜花（菊花、茉莉花、玫瑰花等）经独特的手工艺与现代技术相结合精制而成，产于安徽黄山、福建福鼎等地。

Craft tea, taking the superior tea as raw materials, is refined with dehydrated flowers (chrysanthemum, jasmine, rose, etc.) by combining unique handicraft with modern technology. This kind of tea is usually produced in Huangshan Mountain, Anhui and Fuding, eastern Fujian.

图1.52 工艺造型茶
Craft tea

老茶包

高 21、长 55、宽 50 厘米，重 32 千克

产地：四川省

此茶包为20世纪50年代四川一带的边销茶。中国国际茶文化研究会原副会长、西南茶文化研究中心主任孙前先生在考察川藏茶马古道时，走访了17个县，在四川甘孜与西藏、青海的交界处发现。

茶包由牛皮包装而成，顶部和底部皆裹着成块的牛皮，两边用线牵拉缝合。内装的茶包是由三条竹篾包装的茶包改装而成。这种扁长方形的茶包便于贮存和运输。

太姥山银针

装藏银针重量：103 千克

产地：福建省福鼎市

为纪念中国茶叶博物馆2022年基本陈列提升改造，由方守龙、林逢阳先生亲制太姥山银针。

这缸太姥山银针装藏了2022年采制的白毫银针。银针芽头肥壮、匀齐，挺直如针，满披白毫，色泽如银。汤色杏黄、明亮，滋味醇厚甘甜。

庆祝中国共产党建党100周年·皇室龙团

高35、腹径56厘米，重50千克（100斤）

产地：云南省

龙团采用普洱茶原料，通过手工揉制成瓜形，由普洱茶（贡茶）制作技艺传承人李兴昌、李明泽父子亲制。

普洱贡茶主要产自云南省普洱市宁洱哈尼族彝族自治县。贡茶制作技艺拥有浓厚的历史文化内涵和独特的生产工艺。此茶造型圆润饱满，气势磅礴，形状自然优美，独具一格。

中国茶叶博物馆建馆三十周年纪念·青砖茶圆饼

直径100厘米,净重66.8千克
产地:湖北省

2021年,湖北省赵李桥茶厂有限责任公司为庆祝中国茶叶博物馆建馆三十周年精心制作大茶饼,以优质老青茶为原料,经高温蒸压而成。

青砖茶产于湖北省的蒲圻、咸宁、通山、崇阳等县,色泽青褐,香气纯正,汤色红黄,滋味香浓醇和。

第五章 茶叶品饮与储存
Tea Drinking and Storage

> 茶的冲泡和品饮讲究科学性和艺术性。所谓科学性就是要了解不同茶叶的特点，掌握科学的冲泡技术，使茶叶的内在品质充分表现出来。所谓艺术性就是要根据不同的茶叶选取相宜的茶具，讲究冲泡程序和环境氛围的营造。不良的储存环境，会导致茶叶变质，影响饮用。因此，茶叶的保存也至关重要。
>
> Tea brewing and drinking are scientific and artistic. The so-called "scientificity" means understanding the characteristics of different teas and mastering the scientific brewing skills, so that the inherent quality of teas can be fully displayed. In regards to "artistry", it is to select suitable tea sets according to different teas, and to pay attention to brewing procedures and the creation of environmental atmosphere. Poor storage environment will lead to tea deterioration and affect drinking. Therefore, the preservation of tea is also very important.

一

沏茶用水
Water for Brewing Tea

"水为茶之母"，好茶须有好水配。唐代陆羽有"山水上，江水中，井水下"的择水论断。明代张大复在《梅花草堂笔谈》中说："茶之性发于水，八分之茶，遇十分之水，茶亦十分矣；八分之水，试十分之茶，茶只八分耳。"现代人在古人基础上总结出"清、轻、甘、活、冽"的评水标准。

"Water is the mother of tea." Good tea must go with good water. In the Tang Dynasty, Lu Yu set forth his view on water selection as "Mountain water is the first choice, followed by river water and then well water". Zhang Dafu, a scholar in the Ming Dynasty, stated: "Flavor is brought out by water. If made with superior water, moderate tea tastes quite good. However, superior tea made with moderate water tastes not as good as it should be." On the basis of the ancients, modern people recommend "clear, light, sweet, fresh and icy" water.

● 评水标准
Water Evaluation Standard

现代泡茶用水，一般选用自来水、矿泉水、纯净水等。

Tap water, mineral water, purified water, etc. are generally used for modern tea making.

清：要求水质无色透明、无沉淀物。
Clear: The water should be colorless, transparent and free of sediment.

轻：根据所含钙镁离子的高低，水分硬水和软水。用软水沏茶，色、香、味俱佳。
Light: There are hard water and soft water. Tea made with soft water is excellent in color, fragrance and taste.

甘：指水味甘甜。
Sweet: Sweet water is preferred.

活：要求水"有源有流"，不是静止水。
Fresh: "Active and flowing" water is good for tea making.

冽：指水含口中有清凉感，因而古人十分推崇以雪水煮茶。
Icy: Cool and refreshing water, such as snow highly praised by Chinese ancients, is a good choice.

图1.53 江苏镇江中泠泉
Jiangsu Zhenjiang Zhongling Spring

图1.54 江苏无锡惠山泉
Jiangsu Wuxi Huishan Spring

图1.55 浙江余杭双溪陆羽泉
Zhejiang Yuhang Shuangxi Luyu Spring

图1.56 浙江杭州虎跑泉
Zhejiang Hangzhou Hupao Spring

二

沏茶用具
Utensils for Brewing Tea

"器为茶之父",泡茶要考虑茶叶与茶具材质的特性,两者相互匹配,泡出的茶才能美味和美观兼备。茶具的材质种类繁多,除常见的陶、瓷、玻璃外,还有玉石、竹木、漆器、金属等。

"Utensil is the father of tea." When making tea, one should consider the features of tea and tea set materials. Only when they match each other, can the brewed tea be both delicious and beautiful. There are many kinds of tea sets, including pottery, porcelain, glass, jade, bamboo, lacquer and metal.

● 陶质茶具
Pottery Tea Wares

陶质茶具,以最负盛名的紫砂茶具为例。用紫砂壶泡茶,香味醇,保温性好,无熟汤味,一般用来泡乌龙茶、黑茶等原料成熟度高的茶类。

As a kind of pottery ware, the most famous purple clay tea wares serve an excellent example. Tea made in a purple clay teapot is mellow in flavor, heat-preserving and without the smell of overcooked infusion. It is generally used to make oolong tea, dark tea and other teas.

图1.57 紫砂壶泡法
Purple clay pot method

图1.58 盖碗泡法
Covered bowl method

● 瓷质茶具
Porcelain Tea Wares

瓷质茶具，能很好地反映出茶汤色泽，且传热、保温性适中，能获得较好的色、香、味。瓷质茶具适用广泛，绿茶、红茶、乌龙茶等都可选用。

The color of tea infusion stands out in porcelain tea wares, which conduct and preserve heat moderately, and thus promise charming color, fragrance and taste. Porcelain tea wares are usually used to brew green, black and oolong teas.

● 玻璃茶具
Glass Tea Wares

玻璃茶具，价廉物美、质地透明、导热快，适合冲泡细嫩的名优绿茶及造型花茶。用玻璃茶具冲泡茶叶，便于观察茶叶在水中缓慢舒展、变化的过程，可充分欣赏茶叶的外形。

The glass tea ware is cheap, transparent and fast heat-conducting. It is a great match for delicate famous green tea and shaped scented tea. While brewing tea in glass tea ware, one can readily observe tea slowly stretching and changing in water, and appreciate the appearance of tea.

图1.59 玻璃杯泡法
Glass bubble method

三

茶叶冲泡
Tea Brewing

不同茶类，冲泡方法各不相同。以正确的方式冲泡，才能激发出茶香和茶味。

Brewing method varies based on tea category. Brewing properly is the only way to bring out the aroma and taste of tea.

● 泡茶水温
Water Temperature

水温的高低主要根据茶叶品种而定，如细嫩的绿茶、红茶、黄茶，以85～90℃为宜；乌龙茶、白茶、黑茶、花茶等，宜用100℃的水来冲泡；紧压茶可用煮饮的方式，充分提取茶叶的有效成分。

That water at what temperature can be used for brewing mainly depends on the variety of tea. For delicate green, black, and yellow teas, 85 – 90℃ is appropriate; oolong, white, dark, scented teas, etc., 100℃. Compressed tea can be boiled to fully extract the active ingredients of tea leaves.

● 茶水比例
Ratio of Tea to Water

茶水比例不同，茶汤香气的高低和滋味浓淡各异。冲泡绿茶、红茶、黄茶及花茶的茶水比例为1∶50左右；乌龙茶的茶水比例为1∶20左右。煮饮茶类的茶水比例为1∶80左右。泡茶所用的茶水比例依消费者的嗜好而异。

图1.60 绿茶冲泡
The brewing of green tea

图1.61 红茶冲泡
The brewing of black tea

Tea-water ratio affects the aroma and taste of tea soup. The ratio for green, black, yellow and scented teas brewing is 1:50 or so; oolong tea, 1:20 or so. That for tea boiling is about 1:80. The proportion of tea-water also varies according to consumers' preferences.

● 冲泡次数
Brewing Times

茶叶的冲泡次数与茶叶种类、泡茶水温、茶水比例都有关系,一般绿茶、红茶、花茶等冲泡三次为宜,乌龙茶、黑茶等可多次冲泡品饮。

Times of tea brewing are relevant to the category of tea, water temperature, and tea-water ratio. Generally green tea, black tea, scented tea, etc. can be brewed three times. Oolong tea and dark tea can be brewed and consumed several times.

四

茶叶储存
Tea Storage

影响茶叶储存的主要因素是温度、湿度、氧气、光照等。茶叶的储存，首先要求茶叶含水量低，其次要求储存环境低温、干燥、避光、密封，减少与空气的交换量。

The main factors affecting the storage of tea are temperature, humidity, oxygen, light, etc. First of all, tea having low moisture content can be kept better. Secondly, tea should be kept in cool and dry places. Also it should be sealed and kept out of the sun, reducing the amount of exchange with air.

- 防高温，低温储存
  Keep in Cool Places

茶叶在低温时陈化缓慢，温度高时则品质下降快，宜在低温（0~5℃）环境中储存。

Tea ages slowly at a low temperature and goes bad easily at a high temperature, so it is desirable to store it in cool places at temperatures between 0℃ and 5℃.

- 防潮湿，干燥储存
  Keep in Dry Places

当茶叶中的含水量过高时，茶叶容易陈化和变质。茶叶必须干燥后（含水量在6%以下）进行储存，储存空间的相对湿度最好控制在50%以下。

If moisture content in tea leaves is too high, the leaves are prone to go bad. Only after drying (moisture content below 6%) can tea leaves be stored well, and the relative humidity of the storage space should preferably be controlled at 50% or lower.

图1.62 清 紫砂粉彩八宝纹茶叶罐（中国茶叶博物馆藏）
Doped color tea canister with a eight treasures pattern, Qing Dynasty, China National Tea Museum

- **防氧化，脱氧密封储存**
  Deoxidize and Seal the Tea

茶叶中类脂物质的氧化会产生陈味，氨基酸的氧化可导致鲜味下降，茶多酚氧化会使茶味变淡。因此，茶叶必须进行隔氧密封储存。

Oxidation of lipid substances in tea leaves will produce a stale flavor, and that of amino acids makes tea less fresh. Oxidative poly acids make tea taste lighter. Therefore, tea leaves must be sealed and deoxidized in its storage.

- **防光照，避光储存**
  Keep in Shady Places

光照对茶叶有破坏作用，光线中的红外辐射可转化成热能，使茶叶升温。茶叶在光线的照射下，还会使叶绿素分解褪色，因此包装材料宜选用遮光材质。

Light has a destructive effect on tea leaves. The infrared radiation in the light can convert into heat, causing the leaves to heat up. Besides, direct light will decompose the chlorophyll in tea leaves, so shading materials should be chosen.

- **防吸附，单独储存**
  Keep Tea Separate from Other Items

茶叶本身质地松而孔隙多，具有很强的吸附性能。因此，茶叶必须单独储存，不得用有挥发气味的容器或易吸附其他异味的容器装储茶叶。

Tea leaves are loose in texture and porous, with strong adsorption properties. Therefore, they must be stored separately. Containers with volatile odors and those easily adsorbing unpleasant smells should be avoided in tea storage.

图1.63　晚清 锡质錾花贴画人物纹花形茶叶罐（中国茶叶博物馆藏）
Flower-shaped tin tea canister with figure patterns, the late Qing Dynasty, China National Tea Museum

第六章 茶与健康

Tea and Health

茶叶是一种天然、健康、绿色的饮料。研究表明，茶叶中含有大量的生物活性成分，如茶多酚、茶色素、茶氨酸、茶多糖、生物碱、茶皂素、芳香物质等。从20世纪80年代开始，人们就对茶的活性成分和功能进行了广泛而深入的研究，发现这些成分很多都对人体有营养价值和保健功效，为我们健康饮茶提供了科学依据。随着科学技术的发展，茶的根、茎、叶、花、果等蕴含的营养成分和功效成分不断被开发利用，琳琅满目的茶综合利用产品已然深入到我们生活的方方面面。

Tea is a natural, healthy and green beverage. Studies have shown that tea contains a large number of bioactive components, such as tea polyphenols, tea pigments, theanine, tea polysaccharides, alkaloids, tea saponin, aromatic substances, etc. Extensive and in-depth researches have been conducted on the active components and functions of tea since 1980s, and researchers have found that many of these components have nutritional value and health care effects for human body. With the development of science and technology, the nutritional and functional ingredients contained in the roots, stems, leaves, flowers and fruits of tea have been continuously developed and utilized, and a wide variety of tea comprehensive utilization products have penetrated into all aspects of our daily life.

茶的性味与化学成分
Tea flavors and its chemical components

名称	成分特点	性味	药理作用
绿茶	多酚类物质含量较高，滋味清醇带收敛性	性寒凉，味苦（涩），微甘	清热解毒，抗氧化，提神醒脑
黄茶	多酚类物质发生非酶促转化而减少，游离氨基酸较多	性温，味苦	降浊，调脂，降糖，抗动脉硬化
黑茶	多酚类在微生物的作用下产生复杂氧化作用，形成多聚体和氧化产物；多酚类、氨基酸类减少明显；褐色高聚物增多；芳香类物质增多	性温，味苦涩	降浊，调脂，降糖，抗动脉硬化
白茶	茶多酚、游离氨基酸高，茶氨酸最高，有茶黄素、茶红素，可溶性糖低	性凉，味甘，微苦	清热解毒，抗氧化，提神醒脑
青茶	可溶性糖含量最高，有茶黄素、茶红素，氨基酸低，滋味醇厚爽口，天然花果香浓郁持久，饮后回甘留香	性偏温，味甘，微苦	消食提神，下气健胃，调脂
红茶	多酚类经酶氧化，茶黄素和茶红素最高；多酚类、可溶性糖低，游离氨基酸最低；滋味甜醇、浓厚，具甜香	性温，味甘	和胃散寒，调脂，抗氧化，抗栓

来源：陈宗懋、甄永苏主编《茶叶的保健功能》，北京：科学出版社，2020年。
Source: Chen Zongmao and Zhen Yongsu, *The Healthcare Effect of Tea*, Beijing: Science Press, 2020.

PART II

TRACING THE HISTORY OF TEA

寻踪

第四章 茶为清尚——宋元茶文化
Drinking Tea, an Elegant Fashion: Tea Culture in Song and Yuan Dynasties …… 114

一 北苑贡茶 Beiyuan Tribute Tea …… 115
二 茶事艺文 Tea-Related Art and Literature …… 117
三 点茶斗茶 Diancha (a Method for the Preparation of Tea) and Tea Contest …… 120
四 市井茶坊 Tea Houses in Town and City …… 122
五 辽金茶事 Tea Culture of Liao and Jin Dynasties …… 138
六 再传日本 Re-Spread to Japan …… 144
七 承前启后 Tea Culture in Transitional Period …… 146

第五章 茶韵隽永——明代茶文化
Profound Tea Culture: Tea Culture of the Ming Dynasty …… 150

一 废团改散 Abolishing of Cake Tea and Rise of Loose Tea …… 150
二 瀹饮慢品 Brewing and Sipping Tea …… 151
三 茶书撰著 Tea Books …… 152

第六章 茶意不歇——清代茶文化
Endless Charm of Tea: Tea Culture of the Qing Dynasty …… 162

一 华茶出口 Export of Chinese Tea …… 162
二 茶馆风情 Various Tea Houses …… 168
三 清饮成趣 Tea Drinking …… 168
四 茶庄茶号 Tea Shops and Firms …… 169
五 宫廷茶事 Tea in the Royal Court …… 190

第七章 茶业复兴——近现代茶文化
Revival of Tea Industry: Modern and Contemporary Tea Culture …… 198

一 茶人群像 A Group of Modern Tea Experts …… 198
二 传薪播火 Education of Tea Science …… 200
三 茶叶科技 Tea Science and Technology …… 202

100% 茶渣纤维生态板
100% Eco tea-fibers-based panels

茶保健品
Tea health products

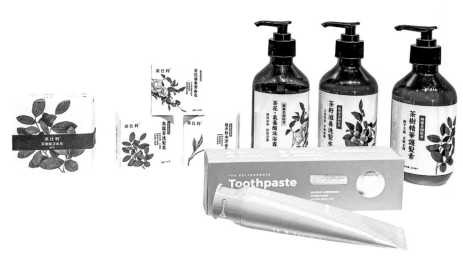

茶日化用品
Tea-based daily chemical products

茶纺织品
Tea-based

茶日用品
Tea-based articles for everyday use

茶日化用品
Tea-based daily chemical products

图1.64 茶综合利用产品
Tea-based products for multiple purposes

茶史

第二篇

第一章 茶生南国——饮茶的起源
Tea Grows in Southern China — Tracing the Origin of Tea Drinking
063

一 茶树原产地 Origin of Tea Trees ⋯⋯ 063

二 饮茶起源说 Origin of Tea Drinking ⋯⋯ 066

三 「茶」的古老称谓 Pre-Tang Terminology of Tea ⋯⋯ 067

第二章 茶风初扬——先秦至魏晋南北朝茶文化
Formative Period of Tea Culture — Tea Culture from Pre-Qin to Wei, Jin, Southern and Northern Dynasties
070

一 考古发现的茶叶 Tea in Archaeological Findings ⋯⋯ 070

二 作为商品的茶叶 Tea as a Commodity ⋯⋯ 073

三 滋味播九区 Popularity of Tea ⋯⋯ 078

第三章 茶事盛景——唐代茶文化
Flourishing Tea Culture — Tea Culture in the Tang Dynasty
082

一 勃兴之因 Causes for the Spread of Tea Drinking ⋯⋯ 083

二 陆羽《茶经》 *The Classic of Tea* by Lu Yu ⋯⋯ 084

三 茶诗风华 Tea Poetry ⋯⋯ 086

四 饼茶制作 Cake Tea Processing ⋯⋯ 088

五 烹煮而饮 Boiling Tea for Drinking ⋯⋯ 088

六 茶马古道 The Old Tea-Horse Road ⋯⋯ 090

七 茶叶外传 Spread of Tea and Tea Culture ⋯⋯ 091

第一章 茶生南国
——饮茶的起源
Tea Grows in Southern China
Tracing the Origin of Tea Drinking

中国的山川大地孕育了无数奇花异树，有嘉木、瑞草、灵草、甘草等雅号美称的茶叶就是其中之一。它的远祖离我们已万分遥远，它的故乡正是我们生活的国度。在此我们将以茶树为起点，缓缓展开茶叶的故事。

Numerous amazing flowers and trees have been nurtured on the land of China, and tea, hailed as *jiamu*, *ruicao*, *lingcao* and *gancao* (meaning good, auspicious or sweet plant) is one of them. It has a very very long history, and its hometown is in China. Starting with the tea tree, here we will slowly unfold the story of tea.

一

茶树原产地
Origin of Tea Trees

茶树的原产地在中国，中国是最早利用茶树、对茶树进行人工栽培的国家。考古学、古植物学、地质学、细胞遗传学、音韵学和语言学等多学科的研究结果均支持这一论断，而丰富的史前遗存则印证了茶与华夏先民之间紧密的联系。

China, the first country to use and cultivate tea trees artificially, is the origin of tea trees. This conclusion is supported by the results of multidisciplinary studies such as archaeology, palaeobotany, geology, cytogenetics, phonology and linguistics, and the rich prehistoric remains verify the close connection between tea and Chinese ancestors.

1. 跨湖桥遗址出土的茶树植物种实
Tea Plant Seed Unearthed from Kuahuqiao Site

2001年发掘的杭州萧山跨湖桥遗址出土了一颗茶树植物种子。该种子为黑褐色，略有炭化现象，呈椭圆形，种脐端略微突出，有裂口，似乎为单室茶果的种子。这颗来自约8000年前的小种子，为茶树利用的起源提供了参考。

A tea plant seed was unearthed from Kuahuqiao Site in Xiaoshan, Hangzhou in 2001. The seed is dark brown and slightly carbonized, with oval appearance, slightly protruding hilum end and cracks. It seems to be the seed of one-seed fruit. This small seed from 8000 years ago provides reference for the origin of tea tree utilization.

2. 河姆渡遗址出土的芽叶纹陶片
Pottery Pieces with Bud and Leaf Pattern Unearthed from Hemudu Site

河姆渡文化第四文化层中出土的陶块正面刻有三片叶子，侧面刻连缀的含芽双叶纹。另外，还有两件陶纺轮和一件刻花陶片上刻划叶纹；一些陶釜的外口沿和肩部刻划树叶纹；一件椭圆六角陶盆的口沿饰有连缀芽叶纹。此处的芽叶纹，颇像茶叶。

A block of pottery unearthed from the fourth cultural layer of Hemudu Culture is carved with three leaves on the front and connective double leaves pattern with buds on the side. In addition, there are two pottery spinning wheels and one pottery piece carved with leaf pattern. The outer edge and shoulder of some pottery pots are carved with leaf pattern. The mouth edge of an oval hexagonal pottery basin is decorated with connective bud-leaf pattern, which looks quite like tea leaves.

图2.1　跨湖桥遗址出土的茶树植物种实
Tea plant seed unearthed from Kuahuqiao Site

图2.2　河姆渡遗址出土的芽叶纹陶片
Pottery piece with bud and leaf pattern unearthed from Hemudu Site

图2.3　河姆渡遗址
Hemudu Site

图2.4　河姆渡遗址出土的芽叶纹陶盆
Pottery basin with bud and leaf pattern unearthed from Hemudu Site

3. 田螺山遗址出土的茶树根
Tea Tree Roots Unearthed from Tianluoshan Site

2004年，在距今约6000年的余姚田螺山人类文化遗址，发掘出两大片树根类植物遗存，树根呈条状、块状或球状。2008年，日本东北大学实验室用显微镜观察田螺山出土树根切片，结果显示这些树木遗存均属山茶科山茶属植物，并认为有可能属于茶树。中国农科院茶叶研究所用树根浸泡液测定茶氨酸的方法进行检测，结果在浸泡液中发现少量茶叶特有的茶氨酸成分。

In 2004, two large remains of tree roots were excavated at Tianluoshan Human Cultural Site in Yuyao, about 6000 years ago, and the roots are strip, block or ball shaped. In 2008, the laboratory of Tohoku University in Japan observed the slices of the tree roots unearthed from Tianluoshan Site with a microscope. The results showed that these tree remains belonged to the *Camellia* plant of Theaceae, and it was believed that they might be tea trees. The Tea Research Institute of Chinese Academy of Agricultural Sciences detected with the method for determination of theanine by root soaking solution, and found a small amount of tea specific theanine in the soaking solution.

图 2.5 田螺山遗址
Tianluoshan Site

图 2.6 田螺山遗址出土的茶树根
Tea tree root unearthed from Tianluoshan Site

图 2.7 田螺山遗址出土的陶壶
Clay pot unearthed from Tianluoshan Site

二

饮茶起源说
Origin of Tea Drinking

目前，饮茶的起源说主要有三种：饮用起源说、食用起源说与药用起源说，其中以药用起源说影响最大。我国现代茶业的奠基人吴觉农先生等人支持药用起源说，曾推断"茶由药用时期发展为饮用时期，是在战国或秦代之后"。顾炎武在《日知录》中列出了一些古文献资料之后，也指出"是知自秦人取蜀而后，始有茗饮之事"。也就是说，至少在战国中期，今天的四川一带已开始饮茶。

There are mainly three different views about how tea was utilized initially. Some argued that it was a drink at the very beginning. Others believed it was used as food. Still others hold the most influential view that tea was treated first as medicine. Wu Juenong, the founder of China's modern tea industry, supported the third view and inferred that, "It is after the Warring States Period or Qin Dynasty when tea was made use of as a beverage instead of medicine". Based on some ancient sources, the famed Qing scholar Gu Yanwu also pointed out in *Rizhilu* (*Record of Daily Study*) that, "Drinking tea didn't happen until the Qin people seized Sichuan". That is to say, at least in the mid-Warring States Period, tea was drunk in today's Sichuan.

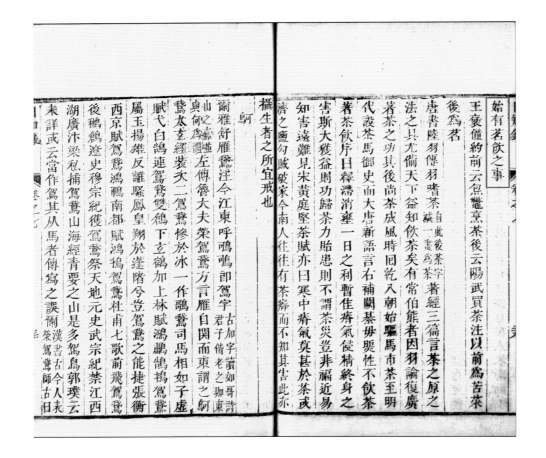

图 2.8　清 顾炎武《日知录》，清康熙三十四年（1695年）吴江潘氏遂初堂本
Rizhilu (*Record of Daily Study*), Gu Yanwu, Qing Dynasty

三 "茶"的古老称谓
Pre-Tang Terminology of Tea

唐陆羽《茶经·一之源》指出，茶的名字，"一曰茶，二曰槚，三曰蔎，四曰茗，五曰荈"。除此之外，唐以前还有搽、檟、葭、荼荈、苦荼、荈诧等叫法和写法。由此可见，虽然早期人们已经种茶饮茶，但依然没有统一的称谓，直到"荼"去一画而出现"茶"。陆羽撰著的《茶经》更推动了"茶"字的流行。

Lu Yu of the Tang Dynasty noted in *The Classic of Tea* that tea is also called "cha (茶), jia (槚), she (蔎), ming (茗) or chuan (荈)". In addition, before the Tang Dynasty, there were other terms for tea such as cha (搽), jia (檟), jia (葭), tuchuan (荼荈), kutu (苦荼) and chuancha (荈诧). This shows that although people had planted and drank tea in the early days, there was still no uniform title for it until "tea (茶)" appeared after one stroke "一" was removed from "tu (荼)". *The Classic of Tea* further promoted the popularity of the word "茶(tea)".

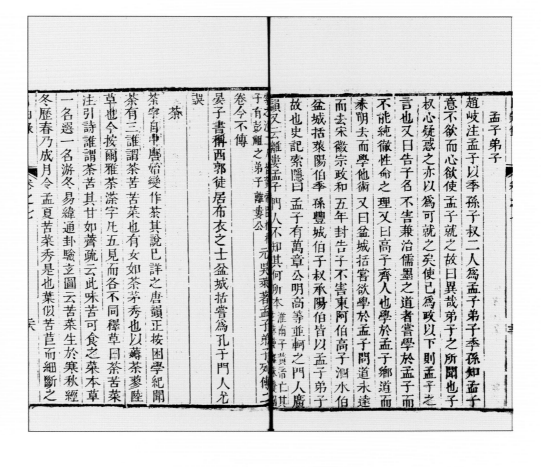

良渚文化 灰陶双鼻壶

高 11.6、口径 7、底径 7.9 厘米

小喇叭形口，两侧置对称的双鼻耳，双耳各穿一小孔，粗长颈，外凸环形腹，喇叭形外撇圈足。

良渚文化 黑陶双鼻壶

高 10.7、口径 8、底径 6.6 厘米

直口稍外撇,两侧置扁平状耳,双耳各穿一小孔,圆腹,平底。器壁薄匀,系快轮拉坯而成。

第二章

茶风初扬
——先秦至魏晋南北朝茶文化

Formative Period of Tea Culture
Tea Culture from Pre-Qin to Wei, Jin, Southern and Northern Dynasties

古巴蜀是中国茶文化的摇篮。公元前316年秦灭蜀之前，茶叶的饮用主要局限于四川一隅。秦人取蜀后，随着国家的统一和各地交流的加强，茶的饮用和茶业逐渐扩展到适合茶树栽培的江南地区，茶文化也由原始、简约而不断充实、丰富。因此，通常以两汉至南北朝时期为中国茶文化的形成期。

The ancient Bashu region is the cradle of Chinese tea culture. Tea drinking was mainly confined to Sichuan before Qin destroyed Shu in 316 BCE. After that, with national unification and increased communication among different regions, tea drinking and tea industry gradually expanded to the Jiangnan area suitable for tea cultivation, and the primitively simple tea culture was enriched. Therefore, the period from the Han to the Southern and Northern Dynasties is usually regarded as the formative period of Chinese tea culture.

一

考古发现的茶叶
Tea in Archaeological Findings

近些年的考古发现，有助于我们进一步了解茶叶利用的时间以及茶文化传播的范围。

1998年在汉景帝陵第15号外藏坑内发现的植物遗存，经鉴定为埋藏了两千余年的古代茶叶。2012年发现于西藏那曲墓葬中的茶叶让茶在两汉时期的传播范围得到了印证。2018年，从山东省邹城市邾国故城遗址战国墓随葬的原始瓷碗中出土了古人煮（泡）后留下的茶渣，将茶文化起源的实物证据提前到战国早期偏早阶段（公元前453—前410年）。

The archaeological discoveries in recent years help us to further understand when tea was used and how far tea culture was spread.

战国 原始瓷竖纹杯

高 8、口径 13、底径 6.5 厘米

口微内敛，弧腹，平底。灰胎，内外施淡青釉，杯外口沿下以篦划装饰竖纹。

The plant remains found in the #15 outer burial pit of the Mausoleum of Han Emperor Jingdi in 1998 were identified as ancient tea buried for more than 2000 years. Tea found in a tomb in Naqu, Tibet in 2012 demonstrates the spread range of tea in the Han Dynasty. In 2018, tea dregs left by ancient people after boiling (brewing) were unearthed from the original porcelain bowl buried in the Warring States Tomb at Zoucheng City, Shandong Province, which advances the physical evidence of the origin of tea culture to the early Warring States period (453–410 BCE).

1. 汉代墓葬中的茶叶
Tea in Han Tomb

汉阳陵是西汉第四位皇帝汉景帝刘启和王皇后的合葬陵墓。2015年，中国科学院地质与地球物理研究所研究人员利用植物微体化石和生物标志物方法，对1998年在汉阳陵帝陵外藏坑发掘出的植物堆积重新进行了科学分析，鉴定出了茶叶遗存。这是目前我国发现的时代较早的茶叶遗存之一。

Liu Qi, the fourth emperor of the Western Han Dynasty, and Empress Wang were buried together in Yangling, the Emperor's Mausoleum. In 2015, researchers from the Institute of Geology and Geophysics, Chinese Academy of Sciences conducted a scientific analysis of the plants excavated from the outer burial pit of Yangling in 1998, and identified the tea remains, which are one of the earliest tea remains found in China.

2. 西藏阿里地区古墓葬中的茶叶
Tea from the Ancient Tomb in Ngari Prefecture, Tibet

2012年，在西藏阿里地区故如甲木寺遗址出土的一件铜器中，发现了疑似茶叶的"食物残体"。经检测，其中包含只有茶叶才同时具有的茶叶植钙体、茶氨酸和咖啡因等。青藏高原本不产茶，这一考古发现说明1800年前左右，青藏高原的先民们已饮茶。

In 2012, the food residue suspected of tea was found in a bronze ware unearthed from the Gurujiamu Temple Site in Ngari Prefecture, Tibet. Detection shows there are tea plant calcium oxalate crystal, theanine, caffeine in the residue. All this is what only tea has at the same time. Since the Qinghai-Tibet Plateau did not produce tea, this finding demonstrates that the ancients living in the Qinghai-Tibet Plateau drank tea about 1800 years ago.

图 2.9 汉景帝阳陵出土的茶叶及分离后的茶叶标本

Tea unearthed from Yangling and tea specimen after separation

扫码了解考古发现中的茶叶

图2.10 阿里地区出土带有茶叶残留的青铜茶具
Bronze utensils with tea residue unearthed in Ngari Prefecture

二

作为商品的茶叶
Tea as a Commodity

汉代，茶开始为更多人所接受，并在流通中逐渐成为商品。较早开始种茶、饮茶的巴蜀地区已经形成了专门交易茶叶的市场。汉代人的饮茶方式大抵是较为原始的羹煮法，即如同煮菜汤一般将茶叶与葱、姜、橘子等混煮。当时也尚未出现专用的烹茶、饮茶器具，往往一器多用。

Tea began to be widely accepted during the Han Dynasty, and gradually became a commodity in circulation. A special market for tea trading was formed in the Bashu area, where tea cultivation and drinking started much earlier. In the Han, tea was prepared by boiling with shallot, ginger, tangerine peels and so on, a primitive way similar to cooking soup. There were no specialized tea boiling and drinking utensils at that time. A utensil often served multiple purposes.

1.《僮约》
Master-Slave Contract

西汉蜀郡资中人王褒想买下一个僮仆，他在《僮约》一文中给这名僮仆定下了繁重苛刻的劳作项目，其中便有"烹茶尽具"及"武阳买茶"两项茶事，可见汉时武阳已有经过加工的商品茶。

Wang Bao from Zizhong County, Shu Prefecture in the Western Han Dynasty enumerates in the contract heavy duties of a slave whom he wishes to purchase. These duties include "brewing tea and preparing utensils" and "buying tea from Wuyang", suggesting there was processed commodity "tea" in Wuyang at that time.

2.《华阳国志》
Record of Huayang State

《华阳国志》是我国现存最早的一部地方史志。这部著作记载了公元4世纪以前西南地区的风土人情和农业资料，其中提到，古巴国"园有芳蒻、香茗"。

Record of Huayang State is the earliest extant local chronicle in China, which records the local customs and agricultural materials of the southwest region before the 4th century. It is mentioned that there are tea leaves in the garden in ancient Ba Kingdom.

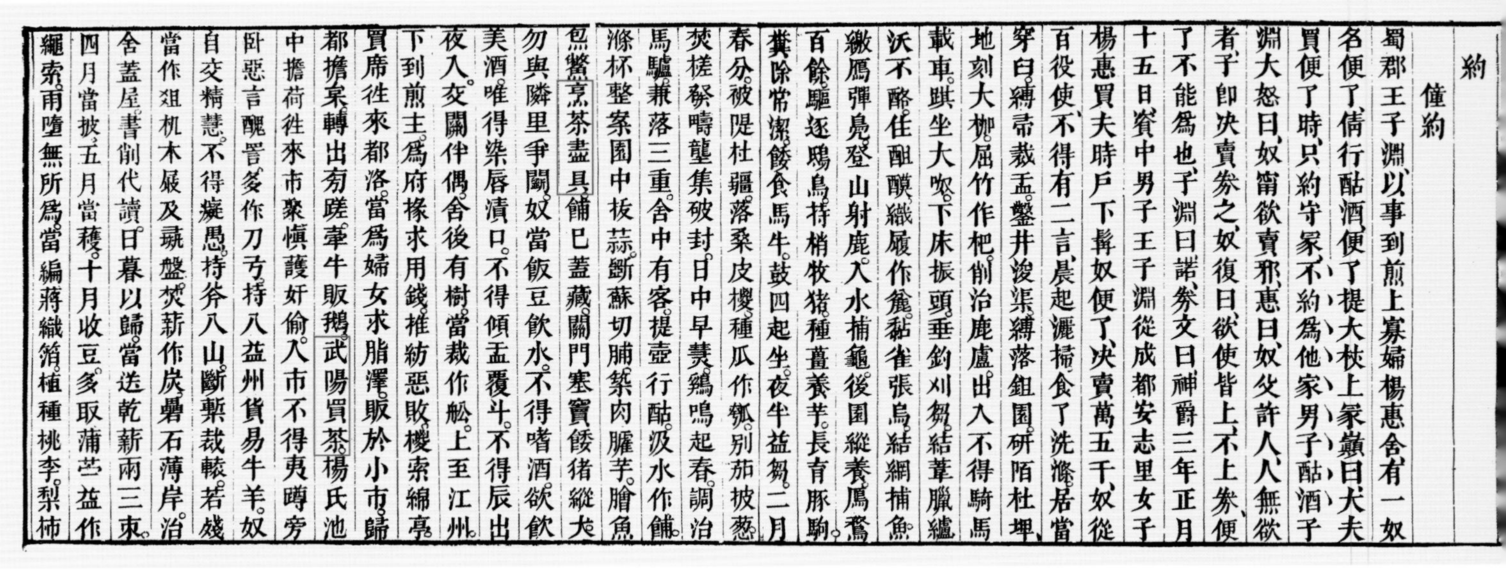

約
僮約
蜀郡王子淵以事到煎上寡婦楊惠舍有一奴
名便了倩行酤酒便了提大杖上冢巔曰大夫
買便了時只約守冢不約爲他家男子酤酒子
淵大怒曰奴寧欲賣邪惠曰奴父許人人無欲
者子卽決賣券之奴復曰欲使皆上不上券便
了不能爲也子淵曰諾券文曰神爵三年正月
十五日資中男子王子淵從成都安志里女子
楊惠買夫時戶下髯奴便了決賈萬五千奴從
百役使不得有二言晨起灑掃食了洗滌居當
穿臼縛帚裁盂鑿井浚渠縛落鉏園研陌杜埤
地刻大枷屈竹作杷削治鹿盧出入不得騎馬
載車踑坐大呶下床振頭垂釣刈芻結葦臘纑
沃不酪住醞釀織履作麤黏雀張鳥結網捕魚
繳雁彈鳧登山射鹿入水捕龜後園縱養雁鶩
百餘驅逐鴟鳥持梢牧猪種薑養芋長育豚駒
糞除常潔餧食馬牛鼓四起坐夜半益芻二月
春分被隄杜疆落桑皮棕種瓜作瓠別茄披葱
焚槎發疇壟集破封日中早䠓雞鳴起舂調治
馬驢兼落三重舍中有客提壺行酤汲水作餔
滌杯整案園中拔蒜斷蘇切脯築肉臛芋膾魚
炰鱉烹茶盡具餔已蓋藏關門塞竇餧猪縱犬
勿與隣里爭鬭奴當飯豆飲水不得嗜酒欲飲
美酒唯得染脣漬口不得傾盂覆斗不得辰出
夜入交關伴偶舍後有樹當裁作船上至江州
下到煎主爲府掾求用錢推紡惡敗棕索綿亭
買席往來都洛當爲婦女求脂澤販於小市歸
都擔枲轉出旁蹉牽牛販鵝武陽買茶楊氏池
中擔荷往來市聚愼護姦偷入市不得夷蹲旁
卧惡言醜詈多作刀弓持入益州貨易牛羊奴
自交精慧不得癡愚持斧入山斷輮裁轅若殘
當作俎机木屐及彘盤焚薪作炭礨石薄岸治
舍蓋屋書削代牘日暮以歸當送乾薪兩三束
四月當披五月當穫十月收豆多取蒲苎益作
繩索雨墮無所爲當編蔣織箔植種桃李梨柿

图2.11　王褒《僮约》
Master-Slave Contract by Wang Bao

五穀牲具六畜桑蠶麻紵魚鹽銅鐵丹漆茶蜜靈龜
巨犀山雞白雉黃潤鮮粉皆納貢之其果實之珍者
樹有荔芰蔓有辛蒟園有芳蒻香茗給客橙（[illegible]當作蕟）
（蕟即橙字）其藥物之異者有巴戟天椒竹木之璝者有
桃支靈壽其名山有塗籍靈臺石書刊山　其民
質直好義土風敦厚有先民之流故其詩曰川崖惟
平其稼多黍旨酒嘉穀可以養父野惟阜丘彼稷多
有嘉穀旨酒可以養母其祭祀之詩曰惟月孟春獺
祭彼崖永言孝思享祀孔嘉彼黍既潔彼犧惟澤蒸
命良辰祖考來格其好古樂道之詩曰日月明明亦
惟其名誰能長生不朽難獲（當有誤）又曰惟德實寶富
貴何常我思古人令問令望而其失在於重遲魯鈍
俗素樸無造次辨麗之氣其屬有濮賨苴共奴獽夷
蜑之蠻　周之仲世雖奉王職與秦楚鄧爲比春
秋魯桓公九年巴子使韓服告楚請與鄧爲好楚子
使道朔將巴客聘鄧鄧南鄙攻而奪其幣巴子怒伐
鄧敗之其後巴師楚師伐申楚子驚巴師　魯莊
公十八年巴伐楚克之　魯文公十六年巴與秦
楚共滅庸　哀公十八年巴人伐楚敗於鄾
是後楚主夏盟秦擅西土巴國分遠故於盟會希

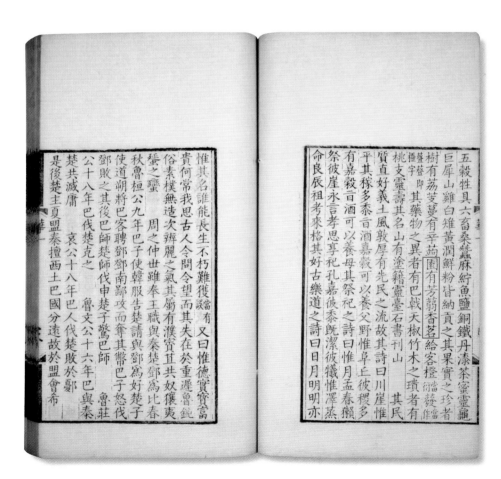

图2.12　《华阳国志·巴志》
The Ba section of *Record of Huayang State*

东汉 原始瓷灶

高 11.7、长 15 厘米

由火膛、烟囱及灶体组合而成，灶体上再承一双耳釜及敛口釜，是东汉时期灶台的真实写照。此系陪葬用的明器。

据文献记载，东汉时期饮茶已在南方四川一带的士人之间流行，不过当时基本上以煮茶为主，称为"茗粥"。

东汉 青铜双龙耳釜

高 10、口径 18.6、底径 16 厘米

敛口，鼓腹下垂，圜底。口沿有双螭龙耳，龙的双角向内，龙首向外张昂，龙身为圆环形，龙尾上翘。造型灵逸，构思巧妙。

汉代王褒《僮约》中已有"烹茶尽具""武阳买茶"之记载。饮茶在司马相如等文人中已较为普遍，当时以煮茶为主，此青铜茶釜系当时重要的煮茶器具之一，汉代的饮茶生活由此可见一斑。

汉 青铜勺

通长 28 厘米

该勺体量较大,柄前端穿孔,串以青铜环,可以悬挂起来,设计巧妙。

汉 青铜火箸

通长 41 厘米

两根箸间有链条相连,是用来夹炭的工具。

东汉 青瓷把杯

高 6.3、口径 8.1、底径 7.1 厘米

直口,筒身,平底,一侧有环形杯把,适合拿捏。胎体较厚,釉青中泛黄,内外施釉,外壁施釉不及底。口沿及腹中部均有两道弦纹,腹部刻划水波纹。

三
溢味播九区
Popularity of Tea

三国魏晋时期，植茶区域与饮茶人群已明显扩大。南北朝时，上层社会以茶迎宾待客、祭祀神灵。北朝宫廷备有茶叶，招待南方来的降臣与使节，而在属于南朝的地区，饮茶的习惯已相当普遍，并开始讲究烹茶时的用水和器具。

这一时期，文人较多地将茶事诉诸笔墨。例如，杜育撰《荈赋》，全面叙述了茶叶的产地、生长、采摘季节以及烹茶用水、茶具、茶汤等。

The tea planting areas and tea drinkers expanded significantly during the Three Kingdoms, Wei and Jin Dynasties. In the Northern and Southern Dynasties, the upper class welcomed guests and worshiped gods with tea. The imperial court of the Northern Dynasty used tea to entertain ministers and envoys from the south. In the areas of the Southern Dynasty, the habit of drinking tea was widespread, and people began to pay attention to water and utensils for tea making.

Many men of letters began to write about tea during this period. For example, Du Yu wrote the *Ode to Tea*, which comprehensively described the origin, growth, picking season of tea, water for tea preparation, tea sets, tea soup, etc.

1. 以茶示俭，以茶养廉
Show Frugality and Nourish Honesty with Tea

两晋、南北朝时期社会风气奢靡，一些人开始用茶与之相抗，赋予茶以俭约、朴素、清淡、廉洁的色彩，丰富了茶的文化精神。东晋桓温"性俭素"，每次宴请仅设七碟茶果待客。陆纳视节俭为"素业"，仅备茶果招待谢安。在此，茶被用来传递自我约束、朴素的生活态度。

During the Jin Dynasties and the Southern and Northern Dynasties, lavish lifestyle was in vogue in the society. Some wise people began to fight against it with tea. The characters of frugality, simplicity, lightness and honesty were given to tea, thus enriching its cultural spirits. Huan Wen of the Eastern Jin Dynasty was frugal, and only seven plates of food were served at his banquet. Lu Na hold "frugality" high and only provided tea and fruit to entertain Xie An. In both cases, tea was used to convey self-restraint and simple attitudes toward life.

2. 采茶作饼，混煮羹饮
Way of Tea Processing and Preparation

三国魏人张揖的《广雅》载曰："荆巴间采茶作饼，叶老者，饼成以米膏出之。欲煮茗饮，先炙令赤色，捣末置瓷器中，以汤浇覆之，用葱、姜、橘子芼之。"它表明，当时已经采用紧压茶叶成饼，以米膏做黏合剂的制茶法。饮用时，将茶叶研磨成屑，再用沸水冲泡或煎煮，还要加入葱、姜等调味品。

It is recorded in *Guangya* by Zhang Yi in the Wei Dynasty of the Three Kingdoms that "In Jingzhou and Bazhou areas, tea leaves are picked to make cake tea. For older leaves, rice paste should be added to make cake tea. To make tea for drinking, one should first roast the cake tea until it turns red, then mash and put it in the ceramic ware, add boiling water to soak it, and finally add shallot, ginger and tangerine peels". It shows that tea at that time was pressed into cakes and rice paste was used as adhesive. When drinking, people would grind tea leaves, then brew or boil them with boiling water, and add shallot, ginger, etc.

图 2.13 东汉—三国 四系印纹"苓"字青瓷罍（湖州市博物馆藏）

Celadon *lei* with the character "苓" and four ears, Eastern Han–Three Kingdoms Period, Huzhou City Museum

东晋 越窑青瓷点褐彩托盘

高 2.5、口径 14、底径 7 厘米

口部微敛，浅弧腹，平底。口沿装饰褐彩，内壁刻划十一瓣莲花，留有五块垫珠垫烧的痕迹，时代特征明显。此系承盏的茶托，为防止茶杯烫手而专门设计。

北朝 青瓷杯

高 6.4、口径 8、底径 3.2 厘米

直口,深腹,高足,足底部微凹。施青釉,釉色偏黄,器内施釉,器外施半釉,有垂釉现象。胎体较厚,胎质较细腻。

南朝 刻莲瓣纹青瓷盏托

高 2.5、口径 10.8、底径 4.5 厘米

敞口,浅弧壁,平底。通体施青釉,盘心有凸起莲心,内壁刻划莲瓣纹,纹饰清晰简练,自然流畅。

第三章 茶事盛景
——唐代茶文化

Flourishing Tea Culture
Tea Culture in the Tang Dynasty

唐代是中国古代文明的黄金时代，也是中国茶文化的黄金时期。中唐以后，在我国大江南北、长城内外，饮茶已蔚然成风；茶叶产区分布广泛，茶之名品不断涌现，丰富人们的饮食生活；文人士子把品茶咏茶作为赏心乐事，以茶为主题的诗文创作成为风尚，更有茶文化专著问世，开茶书撰著之先河，中国茶文化在唐代进入盛世。

The Tang Dynasty is the golden age of ancient Chinese civilization as well as Chinese tea culture. After the mid-Tang, drinking tea was prevalent all over China. Tea producing areas were widely distributed, and famous teas constantly emerged, enriching people's diet and life. Literati thought it was pleasing to drink tea and write for it, and the tea-based literary creation became a fashion. Moreover, the world's first treatise on tea culture was published, creating a precedent for tea book compilation. Chinese tea culture entered its heyday in the Tang Dynasty.

勃兴之因
Causes for the Spread of Tea Drinking

唐代，饮茶之风开始流播于北方，这与佛教禅宗的兴盛及其影响有很大关系，亦与随着大运河通航而出现的水上运输有关。

文人对茶的推崇与宣扬也促使茶文化在唐代蓬勃发展。在文人心目中，茶高洁、雅致，还能助诗兴、发文思。他们为茶作诗撰文，推动了更多人去领略茶中清趣。其中，陆羽与其撰写的《茶经》影响最大。

In the Tang Dynasty, tea drinking began to spread from the south to the north, which is closely connected with the prosperity and influence of Zen Buddhism, as well as the emergence of water transport with the opening of Beijing-Hangzhou Grand Canal.

The booming of tea culture in the Tang also benefited from literati's admiration for and promotion of tea. In the minds of the literati, tea is noble and elegant, and can also stimulate their inspiration and literary thinking. They wrote poems and articles about tea, pushing more people to enjoy tea. Among them, Lu Yu and his *Classic of Tea* were the most influential.

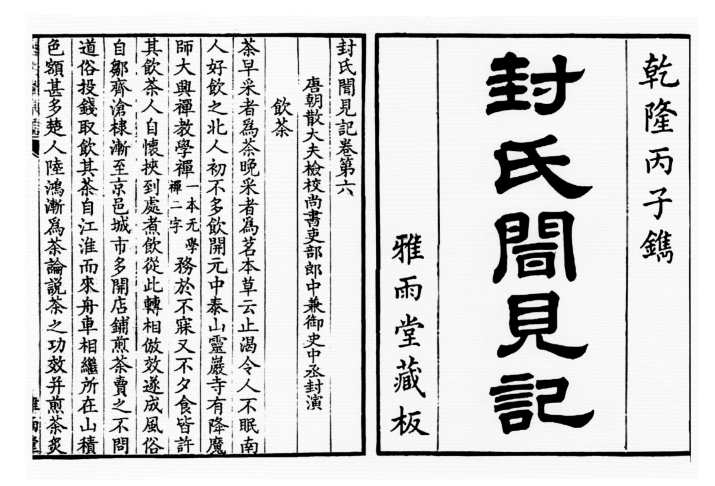

图 2.14　唐 封演《封氏闻见记》
Record of Things Seen and Heard by Mister Feng, Feng Yan, Tang Dynasty

二

陆羽《茶经》
The Classic of Tea by Lu Yu

"自从陆羽生人间，人间相学事春茶。"中唐时陆羽《茶经》的问世，把茶文化推向了一个前所未有的高度。陆羽（约733—804年），字鸿渐，唐复州竟陵（今湖北天门）人。他遍游茶区，考察茶事，前后历时二十多年，写出了世界第一部茶学著作《茶经》，对茶的起源、采制、用具、产区、烹煮、品饮等作了深入细致的研究与总结。《茶经》对当时与后世的茶文化发展产生了深刻影响。因陆羽有功于茶，后人尊奉他为"茶圣""茶神"。

"Since Lu Yu came to the world, people began to learn to make tea." The publication of *The Classic of Tea* by Lu Yu in the mid - Tang raised the tea culture to an unprecedented height. Lu Yu (ca. 733 – 804), known as Hongjian, was born in Jingling (present-day Tianmen, Hubei Province), Fuzhou in the Tang. He spent more than 20 years traveling around the tea areas and investigating tea affairs, finally writing the world's first tea book *The Classic of Tea*. In the text, he made a thorough, detailed study and summary of the origin, picking, making, utensils, production areas, boiling, drinking, etc. of tea. *The Classic of Tea* had a profound impact on the development of tea culture at that time and later. Because of Lu Yu's great contribution to tea, he is honored as "Tea Saint" or "Tea God".

图2.15 五代 白釉陆羽俑（中国国家博物馆藏）
White-glazed figurine of Lu Yu, Five Dynasties, National Museum of China

图2.16 唐 三彩陆羽坐俑（2015年河南巩义市出土）
Tang tricolor glazed pottery sitting figurine of Lu Yu, unearthed in Gongyi City, Henan Province in 2015

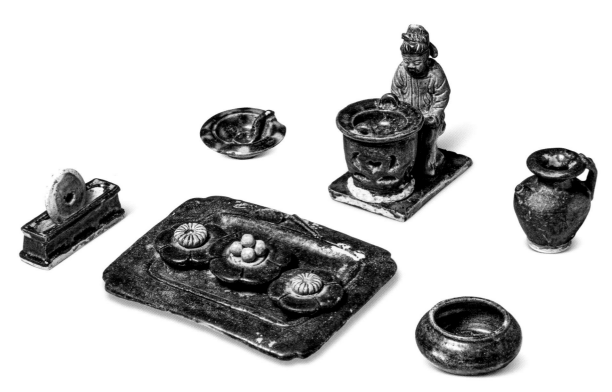

图2.17　唐　三彩茶具组（2015年河南巩义市出土）
Tang tricolor glazed pottery tea set, unearthed in Gongyi City, Henan Province in 2015

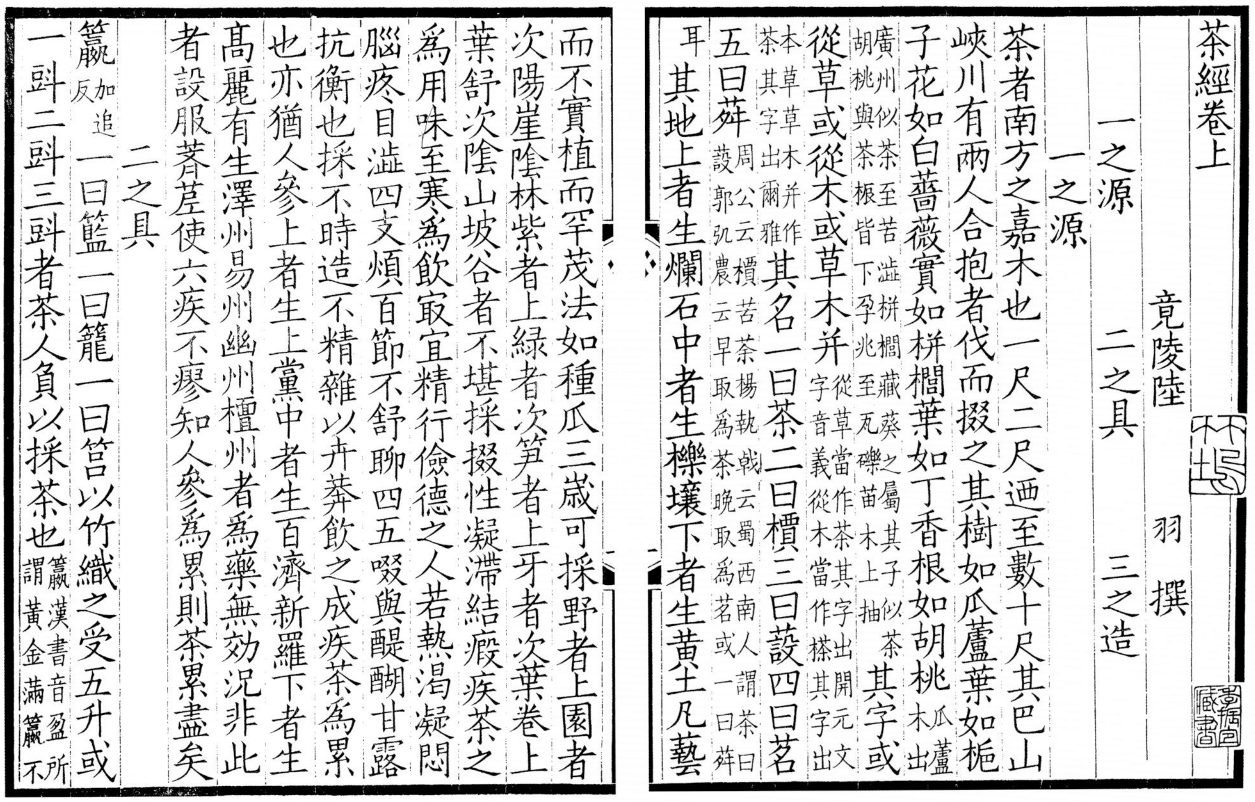

茶經卷上

竟陵陸　羽　撰

一之源　二之具　三之造

一之源

茶者南方之嘉木也一尺二尺迺至數十尺其巴山峽川有兩人合抱者伐而掇之其樹如瓜蘆葉如梔子花如白薔薇實如栟櫚葉如丁香根如胡桃（瓜蘆木出廣州似茶至苦澀栟櫚蒲葵之屬其子似茶胡桃與茶根皆下孕兆至瓦礫苗木上抽）其字或從草或從木或草木并（從草當作茶其字出開元文字音義從木當作梌其字出本草草木并作茶其字出爾雅）其名一曰茶二曰檟三曰蔎四曰茗五曰荈（周公云檟苦茶楊執戟云蜀西南人謂茶曰蔎郭弘農云早取爲茶晚取爲茗或一曰荈耳）其地上者生爛石中者生櫟壤下者生黃土凡藝而不實植而罕茂法如種瓜三歲可採野者上園者次陽崖陰林紫者上綠者次筍者上牙者次葉卷上葉舒次陰山坡谷者不堪採掇性凝滯結瘕疾茶之爲用味至寒爲飲最宜精行儉德之人若熱渴凝悶腦疼目澀四支煩百節不舒聊四五啜與醍醐甘露抗衡也採不時造不精雜以卉莽飲之成疾茶爲累也亦猶人參上者生上黨中者生百濟新羅下者生高麗有生澤州易州幽州檀州者爲藥無効況非此者設服薺苨使六疾不瘳知人參爲累則茶累盡矣

二之具

籝（加追反）一曰籃一曰籠一曰筥以竹織之受五升或一斗二斗三斗者茶人負以採茶也（籝漢書音盈所謂黃金滿籝不所）

图2.18　宋刻本　陆羽《茶经》
The Classic of Tea by Lu Yu, block-printed ed., Song Dynasty

三

茶诗风华
Tea Poetry

唐诗璀璨，光耀千古。在唐代这个茶文化兴盛的重要时期，诗人们纷纷以茶为吟咏的主题，留下了许多值得传诵的诗篇。茶诗的数量，中唐开始时只如涓滴细流，到了唐末已汇成大海。举凡茶叶采摘、焙制、煎煮、品饮，以及名茶、名泉、茶之功效、茶宴等，无不入诗。

The brilliant Tang poetry shines over the ages. In the Tang Dynasty, with flourishing tea culture, numerous poets chanted for tea, leaving many classic poems to the later generations. As for the quantity of poems about tea, what started as a trickle in the mid-Tang became, by the end of the dynasty, a full flood of tea-related verse. The tea poetry elaborates picking, baking, boiling, drinking of tea, as well as famous tea or springs, the efficacy of tea, tea feasts and so on.

- 皎然 Jiaoran

诗僧皎然（约720—约798年）是陆羽的好友与同道中人，一生为茶撰诗多首。最早提出"茶道"一说。

Jiaoran (ca. 720 – ca. 798), a monk-poet, was Lu Yu's friend and kindred spirit. He wrote many poems about tea in his life and first put forward the idea of "tea ceremony".

- 卢仝 Lu Tong

卢仝（约795—835年）也爱茶，曾作《走笔谢孟谏议寄新茶》（又称"七碗茶诗"），其中有诗句描述饮茶的功效，传诵至今。

Lu Tong (ca. 795 – 835), who also loved tea, composed a poem titled *Written in Haste to Thank Censor Meng for His Gift of New Tea* (aka *Seven Cups of Tea*). The famous lines describing the effect of drinking tea have been widely read.

图 2.19 元 钱选《卢仝烹茶图》

Portrait of Lu Tong Boiling Tea by Qian Xuan, Yuan Dynasty

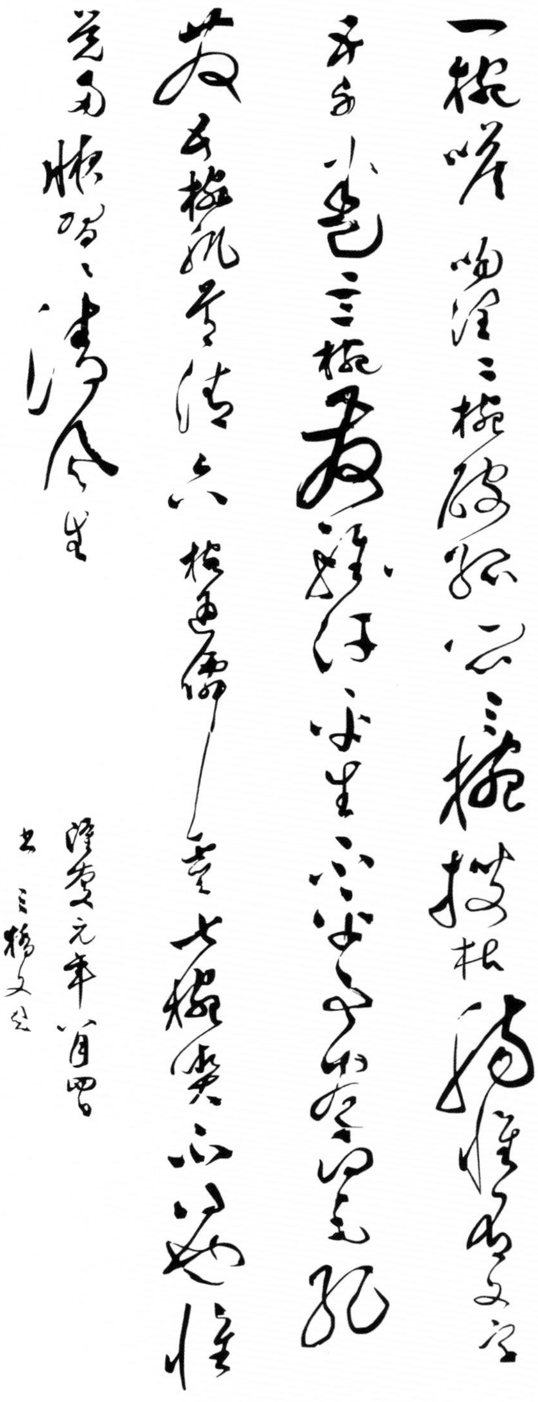

图2.20　七碗茶诗

Seven Cups of Tea

诗句曰:"一碗喉吻润,两碗破孤闷。三碗搜枯肠,唯有文字五千卷。四碗发轻汗,平生不平事,尽向毛孔散。五碗肌骨清,六碗通仙灵。七碗吃不得也,唯觉两腋习习清风生。"

The most famous lines from *Seven Cups of Tea*: "The first bowl moistens my lip and throat. The second bowl banishes my loneliness and melancholy. The third bowl penetrates my withered entrails, finding nothing there except five thousand scrolls of writing. The fourth bowl raises a light perspiration, as all the inequities I have suffered in my life are flushed out through my pores. The fifth bowl purifies my flesh and bones. The sixth bowl allows me to communicate with immortals. The seventh bowl I need not drink. I am only aware of a pure wind rising beneath my two arms."

四
饼茶制作
Cake Tea Processing

唐代茶叶有粗茶、散茶、末茶、饼茶之分，其中饼茶是当时的主流茶品。根据《茶经·三之造》，饼茶的加工需经采、蒸、捣、拍、焙、穿、封七道基本工序，具体即：采茶—蒸茶—捣茶—装模—拍压—出模—列茶（摊晾）—穿孔—烘焙—成穿—封藏。

There were coarse tea, loose tea, tea powder and cake tea in the Tang Dynasty, among which cake tea was the mainstream. According to the "Processing" section of *The Classic of Tea*, seven basic steps are needed for the processing of cake tea: picking, steaming, pounding, patting (putting and pressing tea in mold), baking, threading (a string of tea cakes) and sealing (for storage).

五
烹煮而饮
Boiling Tea for Drinking

唐人将茶叶煮着喝，简言之即先把茶叶碾成末，投入茶䥶（茶釜）中煎煮，最后分酌于茶碗中饮用。与此饮茶方式相对应的茶具有茶䥶（茶釜）、茶铛、茶铫、茶碾、茶臼、茶罗、茶盒、茶则（茶匙）、盐台、茶碗（茶瓯）、盏托等。

Tang people boiled tea for drinking, which means they first ground tea leaves, boiled it in a tea pot (kettle), and finally drank it from tea bowls. The tea sets involved include pot (kettle), *cheng* (pan), *diao* (a small cooking pan), grinder, mortar, sieve, box, spoon, salt container, bowl (tea cup), cup holder, etc.

图 2.21　传唐 阎立本《萧翼赚兰亭图》（局部）
Xiao Yi Trying to Swipe the Lanting Scroll (partial), attributed to Yan Liben, Tang Dynasty

左侧年长者左手持长柄茶铛，置于风炉上，右手持茶箸。右侧侍童手捧黑漆茶托，上坐一白瓷茶碗。侧旁竹质品为"具列"，上置白瓷茶碗及黑漆茶托、碾轮、茶盒等。
The elderly on the left holds a long handle tea pan with left hand and places it on the stove, with tea chopsticks in his right hand. The boy servant on the right holds a black lacquered saucer with a white porcelain bowl on it. The bamboo product on the side is a shelf, on which white porcelain tea bowl, black lacquered saucer, grinder wheel, tea box, etc. are placed.

● 茶鍑 Tea Pots

又称茶釜，系唐代重要的煮茶器具，常与风炉配套使用。通常为敛口，深腹，圜底，或有二耳。鍑（釜）之大口，便于观察水的沸腾状态。

It is an important tea boiling utensil in the Tang Dynasty, also known as the tea kettle, and is often used with a stove. It usually has a converged mouth, deep belly, shallow base, or two ears. The large mouth of the pot (kettle) is convenient for observing the boiling state of water.

● 茶碗 Tea Bowls

又称茶瓯，系典型的唐代茶具。一般分为两类：一类以玉璧底碗为代表，属于陆羽提倡的"口唇不卷，底卷而浅，受半升已下"的器形；另一类为花口，通常作五瓣花形，腹部压印成五棱，圈足稍外撇，多出现于晚唐、五代时期。

Known as tea cup, it is a typical tea utensil in the Tang. Generally it can be divided into two types. One is represented by the bowl with jade disc shaped base, which is described as "having non-curled mouth, shallow bottom and holding basically half a liter of water" by Lu Yu. The other features floral mouth, usually taking the shape of five-petal flowers, with five lines embossed on the body. And the shallow base is slightly outward. Such wares usually appeared in the late Tang and the Five Dynasties.

图 2.22 唐 长沙窑"茶盏子"铭青釉褐彩茶碗（印度尼西亚"黑石"号沉船出水）
Changsha kiln green-glazed tea bowl with brown decoration and the inscription of "tea cup", Tang Dynasty, salvaged from Indonesian "Belitung Shipwreck"

图 2.23 唐 长沙窑"大茶合"铭釉下褐彩茶盒
Changsha kiln under-glaze tea box with brown decoration and the inscription of "big tea box", Tang Dynasty

图 2.24 唐 长沙窑"镇国茶瓶"铭青釉褐彩茶瓶
Changsha kiln green-glazed tea bottle with brown decoration and the inscription of "zhenguo tea bottle", Tang Dynasty

六

茶马古道
The Old Tea-Horse Road

唐贞观十五年（641年），文成公主把茶叶作为陪嫁品带到西藏。"茶马互市"，即用茶叶换回西藏等地良马的交易活动，也始于唐朝，由此逐渐形成了历唐、宋、明、清、民国而不衰的重要商贸通道——茶马古道。茶马古道从横断山脉东侧的云南和四川的茶叶产地出发，穿越横断山脉以及金沙江、澜沧江、怒江、雅砻江，跨过中国最大的两个高原——青藏高原和云贵高原。这是一个庞大而充满艰难险阻的交通网络，是连接汉、藏等多民族的经济文化纽带，也是人类的非凡勇气和超常努力的象征。

扫码了解茶马古道

Princess Wencheng was married to the king of Tibet in 641. Tea, as a dowry, came to the strange land for the first time. The trading of fine teas and quality horses also began in the Tang Dynasty, step by step shaping "the Old Tea-Horse Road" – a significant trading passage that has survived through the Tang, Song, Ming, Qing, and Republic of China periods. Starting from the tea-growing areas in Yunnan and Sichuan (on the east side of Hengduan Mountains), the road winds across the Hengduan Mountains, Jinsha River, Lancang River, Nujiang River and Yalong River, and through the Qinghai-Tibet Plateau and Yunnan-Guizhou Plateau, two greatest plateaus of China. This is a vast and challenging network for commodity, a long-standing link for economic and cultural exchanges among people of different ethnic groups, and a symbol of extraordinary courage and diligence of humanity.

图 2.25　茶马古道上的马蹄窝

Marks left by horses' hooves on the Old Tea-Horse Road

图 2.26　茶马古道上的风雨桥

A bridge on the Old Tea-Horse Road

图2.27 韩国全罗南道宝城茶园
Boseong tea garden in South Jeolla Province, South Korea

七

茶叶外传
Spread of Tea and Tea Culture

中国茶文化源远流长。它不仅润泽、滋养了中国人民的身体与心灵,而且随着来华僧侣与使节的脚步走出了国门,惠益周边的国家与地区。

Chinese tea culture, which has a long history, not only nourished the Chinese people physically and spiritually, but also went abroad and benefited neighboring countries and regions after monks and envoys brought it back home.

1. 传往朝鲜半岛
To Korean Peninsula

《三国史记·新罗本纪》载曰:"茶自善德王有之。"唐文宗时期,新罗使节金大廉把茶籽带回朝鲜半岛,朝鲜自此开始植茶。

According to the *Records of the History of the Three Kingdoms*, "Tea appeared during the reign of Queen Seondeok". During the reign of Emperor Wenzong of the Tang, an envoy of Silla brought tea seeds back to the Korean Peninsula, which marked the beginning of tea cultivation on the peninsula.

2. 传往日本
To Japan

9世纪初，日本僧人最澄、空海入唐求法。回国后，最澄把从天台山带回的茶籽播种在京都比睿山麓的日吉神社，至今那里仍矗立着日吉茶园之碑。空海在归国时也带回了茶籽并献给嵯峨天皇。在日本奈良宇陀郡佛隆寺中，至今仍保存着茶园遗迹和空海带回的石茶碾。

In the early 9th century, Japanese monks Saichō and Kukai came to Tang China for Buddhism learning. After returning home, Saichō planted the tea seeds brought back from Tiantai Mountain in the Hiyoshi Shrine at the foot of Mount Hiei in Kyoto, where still stands the monument of Hiyoshi Tea Garden today. Kukai also brought tea seeds back and offered them to Saga Tennō. The stone tea grinder he brought back and remains of the tea garden are still in Horyu-ji Temple, Nara, Japan.

图2.28　最澄像
Portrait of Saichō

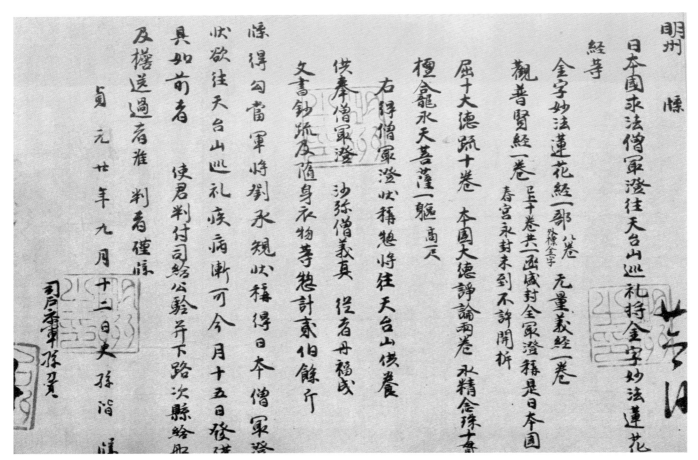

图2.29　最澄大师入唐关牒
Master Saichō's entry certificate to the Tang Dynasty

图 2.30 日本空海法师像及纪念碑
Portrait and monument of Japanese master Kukai

图 2.31 日吉神社御茶园
Imperial tea garden of Hiyoshi Shrine

唐 越窑青釉玉璧底碗

高 3.4、口径 14.6、底径 5.8 厘米

唇口，斜腹，玉璧底。灰胎，器内外施青釉，釉色滋润，器形规整。

唐 邢窑白釉玉璧底碗

高 3.4、口径 14.8、底径 6 厘米

圆唇口，斜腹，玉璧底。胎体细白，施釉光洁。

高 4.5、口径 15、底径 5.5 厘米

口微敛，唇部尖圆，唇沿较厚，断面呈圆弧形，腹圆收，玉璧底。碗心书写"茶埦"两字，"茶"即今"茶"，《说文》："茶，苦茶也。"

长沙窑位于湖南省长沙市望城区铜官镇石渚湖一带，故又称铜官窑，是唐代南方规模较大的青瓷窑场。始烧于初唐，兴盛于中晚唐，终于五代。长沙窑的产品以日常生活器为多，曾生产大量的酒具——大口注子，但这件青釉碗心刻有"茶"字，已十分明确这类碗的用途，可作为长沙窑青釉茶碗中的一个标准器。

唐 越窑青釉茶釜

高 4.3、口径 13.5、底径 5 厘米

敞口，折沿，弧腹，圜底，阔沿双圆耳。整器施青釉，釉色均匀明亮，显示了越窑高超的烧制工艺。

陆羽《茶经》对茶釜有详尽的描述，指出茶釜是方耳有折沿的大口锅，里面光滑，外面粗糙，有利于煎茶和清洗。制作茶釜的材质多样，一般用熟铁铸成，也有洪州（今江西境内）烧制的瓷釜和莱州（今山东境内）产的石釜，更为讲究的则用银制作。

风炉：高 12.3、口径 12、底径 7 厘米
茶釜：口径 13.3 厘米

由风炉和茶釜两部分组成，可拆分。风炉呈筒状，上侈下小，口沿有三小洞，利于通风，一侧开口，以投薪炭烧火，炉门口以下部分出沿，底部圈足外撇；内涩胎，外上姜黄釉，以刻划纹为装饰。釜折沿，浅弧腹，口部有两桥形耳，内施黄釉。风炉和茶釜系重要的煮茶器具。

唐代流行煮茶，又称煎茶。茶饼经过炙、碾、罗后变成茶粉，放入茶釜中煎煮后，入盏饮用。

唐 巩县窑三彩釜

高 10.3、口径 12.4、底径 5.8 厘米

敛口，鼓腹，平底，两侧各有一系。因形如兔子，俗称"兔耳罐"。器内不施釉，外壁施白、绿、赭三色釉，施釉不及底，露胎处呈白色。此罐器形丰满端庄。

高 8.8、通宽 17.7、底径 6.6 厘米

敞口，折沿，深腹，圈足。一侧有匜式流，并带空心横把，横把上装饰弦纹。器内壁施绿釉，外壁上半部分施绿釉，下半部分涩胎，露红色胎质。

唐 黄釉执壶

高 18.1、口径 5.8、底径 7 厘米

撇口,长颈,椭圆形深腹,平底,肩部一侧为流,相对一侧置双带形曲柄,另两侧各有一系。器身施黄色釉,内壁满釉,外壁施釉不及底。

唐 兽流石急须

通高 10.7、通宽 16.4 厘米

附平纽盖，器身敛口，圆鼓腹，腹下渐收，平底。一侧附兽形短流，另一侧置单平把，把下有半圆形系。整器以滑石雕琢而成。该件器物的形制和台湾自然科学博物馆藏的唐代花岗岩茶器组中的侧把壶相似。

唐 长沙窑绿釉茶铛

通高 12.5、口径 16 厘米

敞口,束颈,鼓腹,圜底,下承三足,口沿立两耳。器内施绿釉,器外不施釉。唐代盛行煎茶或煮茶,茶铛系重要的茶器之一。

唐 长沙窑酱釉"张上"印款侧把壶

通高 27、口径 5、底径 8.8 厘米

直颈，圆肩，弧腹修长，足外撇、内凹，口上有盖，盖中部隆起，顶置宝珠纽。壶肩一侧有细长弯流，与之呈直角的一侧置有一斜把，把上有凸起的"张上"二字。壶外壁通体施酱釉。

五代 白釉花口带托盏

盏：高 4.3、口径 12.5、底径 5 厘米
托：高 2.1、口径 7.6、底径 4.7 厘米

由盏和托组成。盏五瓣花口，弧腹，矮圈足。托盘口呈卷荷形，斜壁，矮圈足，中有凸圈以承盏。盏和托均为白胎，釉色白而滋润。

晚唐开始，流行花口盏及托。盏托最早可追溯到东晋时期，当时基本上以圆形茶盘上承碗盏。后随着饮茶兴盛，盏托的形制颇多，茶托有内凹，也有上凸如高台子，盏口有圆形，也有花口等，在实用功能基础上，艺术效果不断加强。

唐 鸟柄铜茶则

长 7.1 厘米

匙面为铲形，两侧上卷。匙柄巧妙地做成展翅的鸟的形状，鸟背部及尾部錾刻出点和线作为装饰，形象地表现了鸟类羽毛的质感。整件器物构思巧妙，制作精致。

唐 花蕾柄铜茶则

长 26 厘米

匙面为叶形，微凹，前后端狭长，匙柄微曲，尾端延伸成一小环，环顶端为莲花的花苞状。整体呈黑色，光可鉴人，线条优美，尾端的花苞精致可爱。

唐 鹊尾柄铜茶则

长 27 厘米

匙面椭圆形,较浅,长柄,柄身为方柱形,后端捶打成扁状。柄身和匙面交界处錾刻有圆珠形纹饰,尾部亦有类似的纹饰。该匙的尾部断裂后被接鹊尾形尾部。

唐 长沙窑茶碾

碾槽高 7、碾长 33.8、轮直径 11.5 厘米

由碾槽和碟轮组成。碾槽略呈四边长方体，内涩胎开槽便于碾轮运动，外壁两侧模印点状及叶纹，施青黄釉。轮素胎无釉，呈圆饼状，中开圆孔以装柄，惜柄失。

唐 长沙窑茶臼

高 3.7、口径 14.7、底径 5 厘米

敞口,弧腹,玉璧底。口沿施酱釉,器内无釉,以篦划纹划出花朵形状图案。茶臼是唐代重要的研磨茶器,与棒杵配合使用。

唐 耀州窑茶臼及研杵

茶臼：高 4.6、口径 15、底径 7.7 厘米

研杵：长 10 厘米

    茶臼为碗形，唇口，斜弧腹，圈足；口沿及外壁施黑釉，施釉不及底，露出土黄色涩胎；内壁以五道平行线，刻划九瓣花朵图案，增加摩擦力，便于研磨。研杵呈土黄色，蘑菇状，与茶臼配套使用。该类研磨器通常用来研磨茶叶及药材。

唐 越窑青釉"荼"字铭瓷盒（残）

直径 8.5 厘米

茶饼碾成茶末后，需有容器盛储，茶盒就是当时盛装茶末的容器。茶盒分两类，尺寸较大类似于捧盒者，系盛装团茶、饼茶之容器；而尺寸较小者则是装盛茶粉末的。卢纶在其诗中写道："三献蓬莱始一尝，日调金鼎阅芳香。贮之玉合才半饼，寄与阿连题数行。"这首唐诗中提及的"玉合"就是装茶饼的容器，所谓玉合（盒）并不是指玉质的盒子，而指越窑青釉盒子的釉色如冰似玉。长沙窑也生产茶盒，其中有褐彩"大茶合"三字的盖盒，证明此类瓷盒系装茶用的容器。

此盖盒虽只留下了下半部分，但此盒底部的"荼"字刻款，却是此类盖盒为茶盒的重要物证。

茶铛：高 8.4、口径 11 厘米

瓷杯（附盖）：杯高 6.7、口径 8、底径 3.9 厘米，盖高 2.9、口径 9.3 厘米

花口带流瓷盏：高 4.9、口径 12.4、底径 4.7 厘米

三足带柄瓷盘：高 3.7、通宽 12.2 厘米

花口瓷盏：高 3.7、口径 10.6、底径 4.2 厘米

茶銚（右）：高 4.5、通宽 14.2 厘米

茶碾（附碾轮）：碾槽高 3、长 11.2、宽 2.5 厘米，轮直径 4 厘米

茶銚（左）：高 2.5、通宽 15.6 厘米

瓷杯（附盖）

三足带柄瓷盘

茶碾（附碾轮）

这套茶具由一茶铛、一带盖瓷杯、一花口带流瓷盏、一三足带柄盘、一花口瓷盏、二茶铫、一茶碾组成。胎质白而细腻，釉色白中泛青，系五代邢窑产品。目前关于茶碾、茶铛、茶铫及茶盏的用途十分清楚，但对于三足带柄盘的功能尚不清楚，有待进一步考证。

在唐代，无论饼茶、散茶皆需碾末煮饮，因此茶碾成为重要茶器。

茶铛和茶釜容易混淆，区别在于器物底部，茶铛器底通常带三足；而茶釜则是折沿，深腹，圜底，双耳或无耳。古人的诗文中有"自携折脚铛，煎芽仍带叶"及"老夫平生喜煮茗，十年烧穿折脚铛"的记载，可见铛是有脚（足）的。

茶铫是有柄有流的烹茶器，一般直接置于风炉之上，通常为圜底或是平底，一侧有流，与流呈直角处装有平柄。

茶盏是品茗时用来盛储茶汤的容器。值得一提的是，此套茶具中还有一带流的花口盏，应是分茶器。晚唐五代时期已经出现了点茶，品茶人数较多时，主人通常会点一大碗茶，然后再分到小盏中分饮。

第四章 茶为清尚

——宋元茶文化

Drinking Tea, an Elegant Fashion

Tea Culture in Song and Yuan Dynasties

茶"兴于唐，盛于宋"，宋代在唐代茶文化的基础上拓宽了它的范围与内容。以贡茶为代表的茶饼制作非常精良。从王公贵族、文人墨客到僧道人士、市井百姓，社会各个阶层无不饮茶。茶风的炽盛，推动了茶馆的进一步发展，使宋代成为茶馆文化兴盛的第一个历史时期。入元后，茶文化虽然不如宋时繁盛，但不绝如缕。

Tea culture "rose in the Tang and flourished in the Song," when the culture developed more extensively. The cake tea, represented by tribute tea, is of excellent workmanship. Tea was consumed at all levels of society, from princes, nobles, literati, writers to monks, Taoists and common people. The ubiquitous tea drinking promoted the further development of tea houses, so that the Song Dynasty became the first historical period in which tea house culture was brilliant. In the Yuan Dynasty, tea culture was not as flourishing as in the Song Dynasty, but it was inexhaustible.

一

北苑贡茶
Beiyuan Tribute Tea

从西周到清末，贡茶贯穿于中国古代社会。唐在浙江湖州顾渚山设立贡茶院，北宋立国后，朝廷派使臣在建安北苑（今福建省建瓯市凤凰山）督造龙凤团茶，"以别庶饮"。根据《北苑别录》，贡茶制作工序为采茶、拣茶、蒸茶、榨茶、研茶、造茶、过黄，工艺十分精细。宋徽宗宣和年间（1119—1125年），北苑贡茶盛极一时。

Beginning in the Western Zhou and ending in the late Qing, the history of tribute tea is quite long. The Tang government set up a tribute tea workshop in Guzhu Mountain, Huzhou, Zhejiang Province. After the founding of the Northern Song Dynasty, the imperial court dispatched envoys to supervise the production of dragon-phoenix cake tea in Beiyuan of Jian'an (present-day Fenghuang Mountain, Jian'ou City, Fujian Province), in order to "distinguish from common people's tea". According to the *Beiyuan Record*, the production of tribute tea includes picking, selecting, steaming, pressing, grinding, making and drying, with very sophisticated craftsmanship. During the Xuanhe reign period (1119–1125) of Song Emperor Huizong, Beiyuan tribute tea was in full flourish.

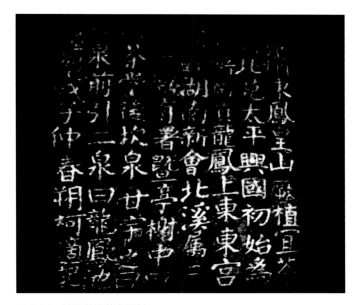

图2.32　建瓯林垅茶事石刻
Linlong stone inscription on tea, Jian'ou

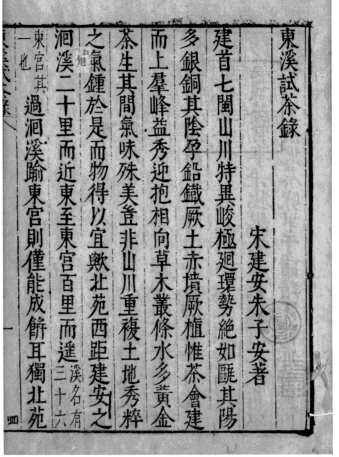

图2.33　宋子安《东溪试茶录》
Record of Tea Tasting in Dongxi by Song Zi'an

万春银叶
银模 银圈
两尖径二寸二分

长寿玉圭
银模 银圈
直长三寸

太平嘉瑞
银模 铜圈
径一寸五分

龙苑报春
银模 铜圈
径一寸七分

宜年宝玉
银模 银圈
直长三寸

南山应瑞
银模 银圈
方一寸八分

图 2.34 北苑贡茶线描图及仿制品

Illustrations and modern replicas of a few Beiyuan tribute teas

线描图出自《宣和北苑贡茶录》。该书由北宋熊蕃撰，熊克增补，记载了 50 余种贡茶的名称，其中大多数创制于大观至宣和年间（1107—1125 年）。熊蕃子熊克于绍兴二十八年（1158 年）摄事北苑，因见其父所作茶录仅著贡茶名号，而无形制，于是绘图 38 幅，期使"览之者无遗恨焉"。熊克增补的北苑贡茶茶模线描图名称、尺寸、材质俱全，使后人得以了解宋代贡茶的形制。

These illustrations are taken from *Record of Beiyuan Tribute Tea of the Xuanhe Period*, which was authored by Xiong Fan and supplemented by Xiong Ke in the Northern Song Dynasty. The names of over 50 tribute teas were recorded, and most of them were created during the periods from Daguan to Xuanhe. Xiong Ke, the son of Xiong Fan, served at Beiyuan in the 28th year of Shaoxing (1158). Seeing that his father included in his text only names of the tribute tea, but no shapes, he provided 38 pictures, hoping that "those who see them will have no regret". The illustrations of Beiyuan tribute tea molds added by Xiong Ke contain the name, size and material, so that later generations can get a good knowledge of tribute tea in the Song Dynasty.

二

茶事艺文
Tea-Related Art and Literature

宋代茶叶生产空前发展，茗饮之风更胜往昔。雅好品茗论茶的文人士大夫往往不惜泼墨挥毫，用诗词书画记录各种茶事活动，抒写他们对茶的情怀。许多文人均有佳作传世，如范仲淹、欧阳修、蔡襄、苏东坡、李公麟、米芾、赵令畤、陆游等，使茶文化更具魅力。

The Song Dynasty witnessed the unparalleled development of tea production, and tea drinking enjoyed more popularity. Literati and scholar-officials loved drinking tea and often recorded various tea activities with poetry, calligraphy and painting, to express their feelings for tea. Many scholars, such as Fan Zhongyan, Ouyang Xiu, Cai Xiang, Su Dongpo, Li Gonglin, Mi Fu, Zhao Lingzhi and Lu You, left masterpieces, making tea culture more charming.

1. 赵令畤《赐茶帖》
Tea Bestowing Calligraphy by Zhao Lingzhi

《赐茶帖》为行书，9行57字，内有"比拜上恩赐茶，分一饼可奉尊堂"句。北苑贡茶尽归皇室所有，皇帝为了表示皇恩浩荡和爱才之心，也会将贡茶赏赐给文人士大夫。对后者而言，获赐贡茶是一项殊荣。受赐者对茶或珍藏，或自品，或与友朋分享，或孝敬父母。

Tea Bestowing Calligraphy, running script, has 9 lines and 57 characters, and contains the sentence "the emperor grants tribute tea and one cake tea can be used to honor parents". The tribute tea in Beiyuan belonged to the royal family, but was also bestowed by emperors to ministers to show their grace and favor. The recipients would see it great honor to be bestowed tribute tea. They might treasure the tea, drink it themselves, share it with friends, or honor parents with it.

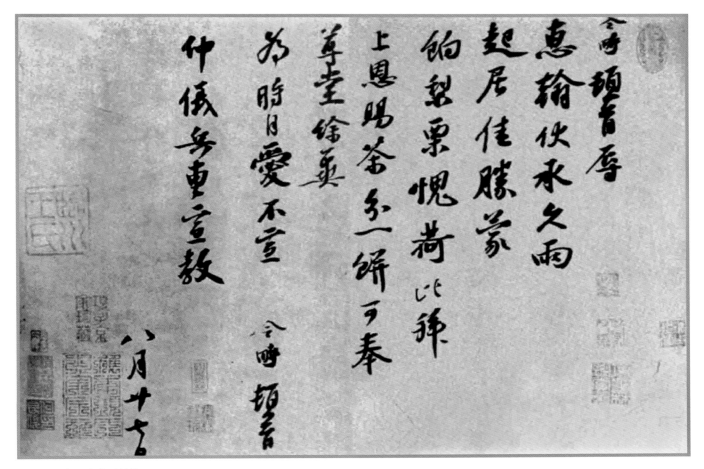

图 2.35　赵令畤《赐茶帖》
Tea Bestowing Calligraphy by Zhao Lingzhi

2. 蔡襄《北苑十咏》
Odes to Beiyuan by Cai Xiang

蔡襄精通茗理，对宋代的茶叶生产及茶文化有突出的影响。他任福建路转运使时在北苑督造贡茶，故而熟谙茶叶的生产环境和制作工艺等，这一点清晰地反映在他的《北苑十咏》中。

Cai Xiang, proficient in tea theory, had a prominent effect on tea production and culture in the Song. While holding the position of commissioner in Fujian, he supervised the production of tribute tea in Beiyuan, so he was familiar with the work, which is reflected in his *Odes to Beiyuan*.

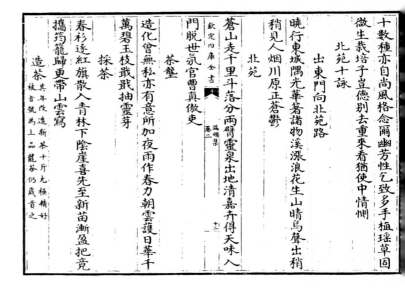

图 2.36 《北苑十咏》
Odes to Beiyuan

3. 苏轼《啜茶帖》
Invitation to Sip Tea by Su Shi

苏轼的书法作品中，关于茶的内容亦不少，《啜茶帖》是其中之一。《啜茶帖》又名《致道源帖》，行书，4行22字。释文："道源无事，只今可能枉顾啜茶否？有少事须至面白，孟坚必已好安也。轼上，恕草草。"书此帖时，苏轼已被贬谪黄州。帖中苏轼邀友人啜茶，茶在其生活中的重要性可见一斑。

Su Shi often included tea in his calligraphic works, and *Invitation to Sip Tea* is one of them. *Invitation to Sip Tea*, also known as *Invitation Extended to Daoyuan*, running script, has 4 lines and 22 characters. It says, "Daoyuan, would you please come over for tea if you are free? I want to have a face-to-face talk with you." By the time Su wrote this invitation, he had been banished to Huangzhou. His inviting friend for tea showed the importance of tea in his life.

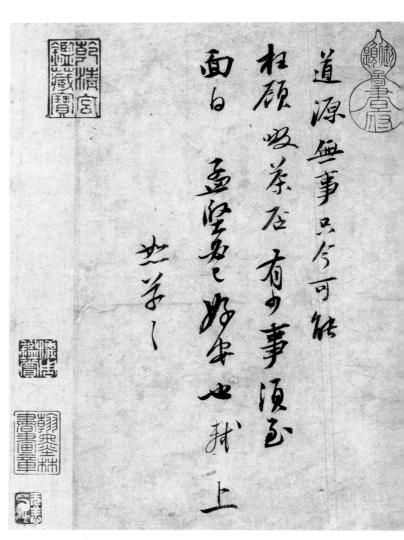

图 2.37 《啜茶帖》
Invitation to Sip Tea

4. 米芾《苕溪诗帖》
Tiaoxi Poetry Calligraphy by Mi Fu

米芾的《苕溪诗帖》涉及茶事，饶具意蕴。诗中记述曰："半岁依修竹，三时看好花。懒倾惠泉酒，点尽壑源茶。主席多同好，群峰伴不哗。朝来还蠹简，便起故巢嗟。"壑源茶产于今福建建瓯壑源一带，苏轼、黄庭坚皆有诗称赞壑源茶。

Tea, specifically Heyuan tea is mentioned in this piece of artwork by Mi Fu. It says that "we are reluctant to pour Huiquan wine, but make a cup after cup of Heyuan tea". Heyuan tea is produced in Heyuan area of Jian'ou, Fujian Province, and is also praised in poems written by Su Shi and Huang Tingjian.

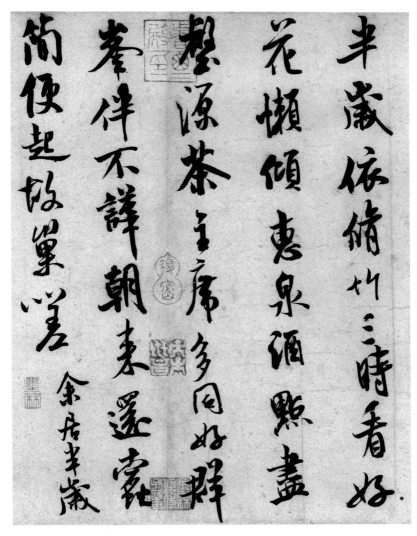

图2.38 《苕溪诗帖》
Tiaoxi Poetry Calligraphy

三

点茶斗茶
Diancha (a Method for the Preparation of Tea) and Tea Contest

宋代饮茶方式有点茶、煎茶。所谓点茶，指将末茶置于茶盏内，先用茶瓶注汤调膏，然后不停用茶匙或茶筅击拂，令茶汤出现乳花。点茶的技艺越高，汤花停留在盏壁内沿的时间越长（名为“咬盏”）。汤花的色泽以纯白为上。宋人还爱斗茶，这是一种集竞技性、娱乐性和艺术性于一体的活动。斗茶讲究茶、水、器，更要有高超的点茶技艺，四者缺一不可。

In the Song, *diancha* and boiling tea were the two methods in which tea was prepared. The former described the way of pouring a thin stream of hot water into tea powder in a tea cup, making the powder into pastry and then whisking with teaspoon or whisk, so that milky bubbles might appear on the surface of the tea soup. The higher the *diancha* skills, the longer the bubbles stay (*yaozhan*, literally the bubbles "bite the cup"). And the pure white soup bubble is the best. The Song people also loved tea contest, which was an activity integrating competitiveness, entertainment and artistry. Tea contest emphasizes the quality of tea, water and utensils, and also requires superb *diancha* skills.

在爱茶人中，宋徽宗赵佶是比较特殊的一位。他以一国之君和品茶大家的身份撰成《大观茶论》一书，总结了北宋茶叶种植、制作与品饮的知识。全书共二十篇，其中“点”讲述点茶要领最为详细。

As the ruler of a country and a master of tea, Zhao Ji, Emperor Huizong of the Song Dynasty, was a special tea lover. He is also the writer of *Treatise on Tea from the*

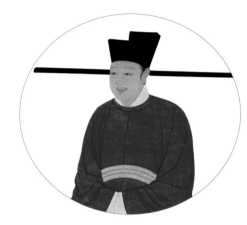

图2.39 宋徽宗像
Portrait of Emperor Huizong

大觀茶論
宋 徽宗
嘗謂首地而倒生所以供人求者其類不一穀粟之
于饑絲枲之于寒雖庸人孺子皆知常須而日用不
以時歲之舒迫而可以興廢也至若茶之爲物擅甌
閩之秀氣鍾山川之靈禀祛襟滌滯致清導和則非
庸人孺子可得而知矣中澹間潔韻高致靜則非遑
遽之時可得而好尚矣本朝之興歲修建溪之貢龍
團鳳餅名冠天下而壑源之品亦自此而盛延及于
大觀茶論 一
今百廢俱舉海内晏然垂拱密勿幸致無爲縉紳之
士韋布之流沐浴膏澤薰陶德化盛以雅尚相推從
事茗飲故近歲以來采擇之精製作之工品第之勝
烹點之妙莫不盛造其極且物之興廢固自有時然
亦係乎時之汙隆時或遑遽人懷勞悴則向所謂常
須而日用猶且汲汲營求惟恐不獲飲茶何暇議哉
世既累洽人恬物熙則常須而日用者固久厭飫狼
籍而天下之士勵志清白競爲閒暇修索之玩莫不
碎玉鏘金啜英咀華較筐篋之精爭鑒裁之別雖下
士于此時不以蓄茶爲羞可謂盛世之清尚也嗚呼
至治之世豈惟人得以盡其材而草木之靈者亦得
以盡其用矣偶因暇日研究精微所得之妙後人有
不自知爲利害者敘本末列于二十篇號曰茶論
地產
植產之地崖必陽圃必陰蓋石之性寒其葉抑以瘠
其味疏以薄必資陽和以發之土之性敷其葉疏以
暴其味強以肆必資陰蔭以節之今圃家皆植木以資茶之陰
陽相濟則茶之滋長得其宜

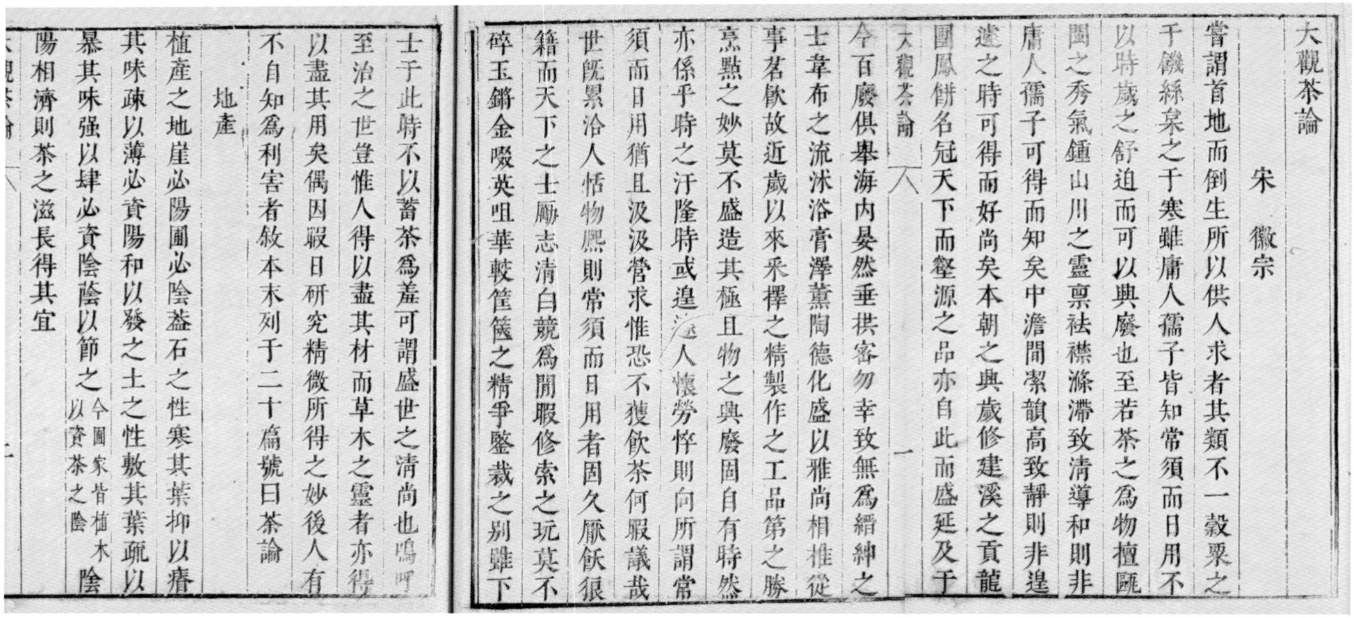

图2.40 《大观茶论》
Treatise on Tea from the Daguan Reign Period

Daguan Reign Period, which summarized the knowledge of tea planting, production and drinking of the day. The text contains twenty sections, and in the section of "Dian" how to *diancha* is elaborated.

• 茶具"十二先生"
Tea Utensils "Mr. Twelve"

南宋审安老人在《茶具图赞》一书中，用白描手法绘制了十二件茶具，并根据每一件茶具的材质、形制和功能，按宋时官制——冠以官职，赐以姓名、字号，贴切传神，生动有趣，令人仿佛面对十二位个性鲜明的人物。

In the book *Illustrated Record of Tea Utensils* by Shen'an Laoren (a pseudonym) of the Southern Song Dynasty, twelve tea utensils were drawn and entitled with official posts according to the material, shape and function of each item. Names and courtesy names were given them too as if they were twelve distinctive and real persons.

图2.41 茶具"十二先生"
Tea utensils "Mr. Twelve"

四

市井茶坊
Tea Houses in Town and City

茶馆的雏形在唐代中叶以前已经出现。唐宋时期习惯称茶馆为茶肆、茶坊、茶楼等。宋代茶坊十分兴盛，是当时经济与文化高度繁荣的产物，也是世俗生活的典型场景。茶坊不仅数量多、经营范围广，而且功能多元，尤其是在北宋都城汴京（今河南开封）和南宋都城临安（今浙江杭州）两地。

The rudiment of tea houses appeared before the mid-Tang. They used to be called tea shops, tea rooms, tea buildings, etc. in the Tang and Song Dynasties. As economy and culture prospered in the Song, the typical space for secular life, tea houses flourished. A large number of tea houses run a wide range of business and also had multiple functions, especially in Bianjing (present-day Kaifeng, Henan), the capital of the Northern Song, and Lin'an (present-day Hangzhou, Zhejiang), the capital of the Southern Song.

1.《清明上河图》中的茶坊
Tea Houses in the *Riverside Scenes at Qingming Festival*

在北宋画家张择端的《清明上河图》中，汴河两岸有赶集人驻足茶坊饮茶歇息，可见京师汴梁茶风之浓郁。

In the *Riverside Scenes at Qingming Festival* by Zhang Zeduan, a painter in the Northern Song Dynasty, on both sides of the Bian River people going to market stop at the tea house to drink tea and rest, showing the strong tea-drinking atmosphere in the capital Bianliang.

图 2.42《清明上河图》之"饮子店"
Drinks shop in the *Riverside Scenes at Qingming Festival*

2. 宋人笔记中的茶坊

Tea Houses in the Song People's Accounts

宋代茶肆盛况空前。宋人吴自牧《梦粱录》第十七卷提到，茶肆“列花架，安顿奇松异桧等物于其上”。耐得翁《都城纪胜》中记载，宋代茶楼中多有都人子弟聚会，或习学乐器。

Tea houses were unprecedentedly prosperous in the Song Dynasty. It is mentioned in the 17th volume of *Mengliang Records* by Wu Zimu of the Song Dynasty that "wondrous pines, cypress and plants are placed on the shelf" in some tea houses. According to the *Records of the Capital* by Naideweng, many young people get together or learn musical instruments in tea houses.

夢粱錄卷十六
錢塘 吴自牧 著
茶肆
汴京熟食店張挂名畫所以勾引觀者留連食客今杭
城茶肆亦如之插四時花挂名人畫裝點店面四時賣
奇茶異湯冬月添賣七寶擂茶饊子葱茶或賣鹽豉湯
暑天添賣雪泡梅花酒或縮脾飲暑藥之屬向紹興年
間賣梅花酒之肆以鼓樂吹梅花引曲破賣之用銀盂
杓盞子亦如酒肆論一角二角今之茶肆列花架安頓
奇松異檜等物于其上裝飾店面敲打響盞歌賣止用
瓷盞漆託供賣則無銀盂物也夜市于大街有車擔設
浮鋪點茶湯以便遊觀之人大凡茶樓多有富室子弟
諸司下直等人會聚習學樂器上教曲賺之類謂之掛
牌兒人情茶肆本非以點茶湯爲業但將此爲由多覓
茶金耳又有茶肆專是五奴打聚處亦有諸行借工賣
伎人會聚行老謂之市頭大街有三五家開茶肆樓上
專安著妓女名曰花茶坊如市西坊南潘節幹俞七郎
茶坊保佑坊北朱骷髏茶坊太平坊郭四郎茶坊太平
坊北首張七相幹茶坊蓋此五處多有炒鬧非君子駐
足之地也更有張賣麵店隔壁黄尖嘴蹴毬茶坊又中

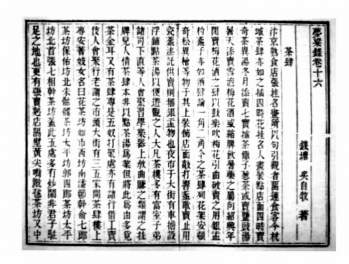

图2.43 南宋 吴自牧《梦粱录》

Mengliang Records by Wu Zimu, Southern Song Dynasty

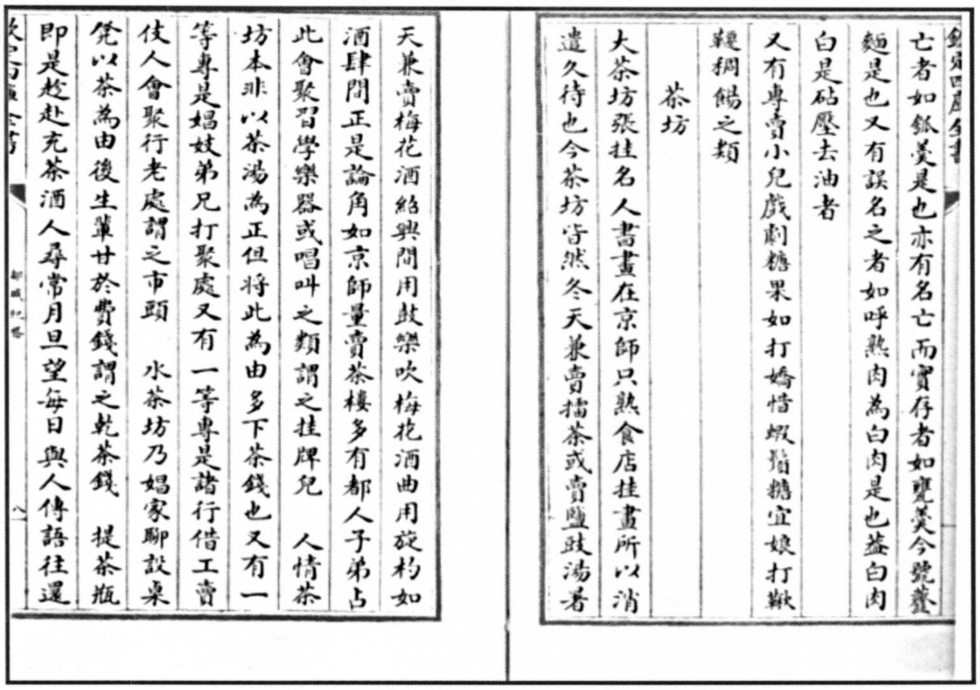
亡者如鈉羹是也亦有名亡而實存者如覽羹今號羹
麵是也又有誤名之者如呼熟肉爲白肉是也蓋白肉
白是砧壓去油者
又有專賣小兒戲劇糖果如打嬌惜蝦鬚糖宜娘打鞦
韆稠餳之類
茶坊
大茶坊張挂名人書畫在京師只熟食店挂畫所以消
遣久待也今茶坊皆然冬天兼賣擂茶或賣鹽豉湯暑
天兼賣梅花酒紹興間用鼓樂吹梅花酒曲用旋杓如
酒肆間正是論角如京師量賣茶樓多有都人子弟占
此會聚習學樂器或唱叫之類謂之挂牌兒 人情茶
坊本非以茶湯爲正但將此爲由多下茶錢也又有一
等專是娼妓弟兄打聚處又有一等專是諸行借工賣
伎人會聚行老處謂之市頭 水茶坊乃娼家聊設桌
凳以茶爲由後生輩甘於費錢謂之乾茶錢 提茶瓶
即是趁赴充茶酒人尋常月旦望每日與人傳語往還

图2.44 南宋 耐得翁《都城纪胜》

Records of the Capital by Naideweng, Southern Song Dynasty

宋 素胎瓷茶臼

高 6.7、口径 15.7、底径 7.5 厘米

呈钵形，一侧带流口。内外均涩胎无釉，内壁以篦划纹划出十组纵横的网格，粗糙的表面可强化研磨的效果。

宋代无论团饼茶或散茶，皆需碾末点茶，茶臼、茶磨、茶碾、茶研均是重要的茶具。

高 8.4、长 32.5、宽 5.5 厘米

碾槽呈船形，如扁舟，下部两侧有支架，中间峻深，以承碾轮。此茶碾为铜质，较少见。

通高 7.5、口径 7.8、底径 4.8 厘米

直口,筒形腹,矮圈足,带盖。灰胎,器外施白釉,器内施透明釉。这类带盖瓷罐有可能也是盛放末茶的容器。

高 24.3、口径 12、底径 8.5 厘米

大侈口，细长颈，弧肩，长圆形腹至胫部渐收，矮圈足。壶口与壶肩之间有扁条形执柄，与之相对的一侧为一长流，流口稍低于壶口部。双线把壶身分为六部分，内暗刻云气纹。

宋代的执壶造型一改唐代的敦朴丰满形象，显得挺拔秀气。

执壶，宋代人通常称之为汤瓶，审安老人给它取了一个更加优雅的名称——汤提点。汤提点是点茶必不可少的茶具之一。中唐时以煎茶为主，投茶末入茶鍑中煎煮，无汤瓶之说。至晚唐、五代时期，点茶开始出现，汤瓶也就应运而生了。罗大经《鹤林玉露》中说："近世瀹茶，鲜以鼎镬，用瓶煮水。"

点茶关键在于汤瓶的流嘴，"注汤利害，独瓶之口嘴而已。嘴之口欲大面宛直，则注汤力紧而不散；嘴之末欲圆小而峻峭，则用汤有节而不滴沥。盖汤力紧则发速，有节而不滴沥，则茶面不破"。此龙泉窑青釉汤瓶基本符合当时点茶的标准。

高 6.9、口径 12.5、底径 4.4 厘米

敞口，深弧腹，小圈足。黑胎较厚，内外施黑釉，在高温作用下，茶盏内外釉面氧化铁结晶而析出丝丝兔毫般的效果。

建窑位于福建省建阳县（今南平市建阳区）水吉镇，以生产黑釉盏而闻名。根据兔毫盏色泽的微妙不同，又分为"金兔毫""银兔毫"和"黄兔毫"。建窑黑釉盏一般胎体较厚，从造型上看，以敛口和敞口两种为多，无论哪种造型，为了点茶的需要，其盏壁都很陡。盏底深利于发茶，盏底宽则便于茶筅搅拌击拂，胎厚则茶不易冷却。

宋 景德镇窑青白釉刻团花纹盏

高 6、口径 16.9、底径 5.4 厘米

撇口，斜腹，小圈足。碗心划刻团花纹，内壁刻一周云头纹，线条生动流畅，格调清雅。口沿涩胎无釉，胎体薄而坚致，内外施青白釉，釉色整体偏青色，秀润青翠，更显精美。

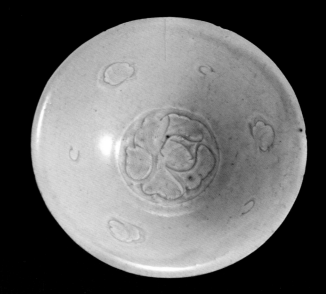

宋 耀州窑青釉模印菊花纹盏

高 5.1、口径 11.3、底径 3.5 厘米

敞口，深弧壁，圈足。内外施青釉，釉色青中泛黄，足边无釉，圈足有窑粘。内壁模印菊花纹，外壁采用剔刻放射性线条的方式表现菊瓣图案。

宋 耀州窑青釉斗笠盏

高 4.6、口径 10.3、底径 3 厘米

唇口，斜深腹，小圈足，近圈足处有窑粘。盏形制规整，形似斗笠，小巧精致。通体施青釉，釉色青翠灵秀，质地致密醇厚。

宋代因点茶需要，茶盏以撇口、斜直壁、尖底小圈足为宜，这样便于茶筅击拂和观赏千奇百态的汤花，因此"斗笠盏"成为宋代南北窑场烧制的热门产品之一。

宋 龙泉窑青釉斗笠盏

高 5.4、口径 12.8、底径 3 厘米

敞口，斜直壁，圈足。内外施青釉，足边不施釉，釉色青翠莹润。口沿稍加金缮。

高 5.2、口径 11.3、底径 3.5 厘米

撇口，斜壁，小圈足。内壁满釉，外壁施釉至足胫部，釉层较薄，呈褐色，整体釉面黑亮，釉色莹润，浑厚简朴。底足处露黄白色胎，坚硬而细润。

吉州窑位于江西吉安永和镇一带，自五代至元代烧制瓷器，又以宋元最为兴盛，极富民间特色。

高 6.2、口径 15.6、底径 4.7 厘米

敞口，弧腹，矮圈足。通体施黑釉，釉面滋润，色彩绚丽。

玳瑁釉是吉州窑的特色产品，瓷器坯体用含铁量较少的瓷土做成，生坯挂釉，入窑烧后再挂一次膨胀系数不同的釉，重烧一次。由于釉层的流动、密集、填缝，便在黑色中形成玳瑁状的斑点，故称玳瑁釉。

高 6.7、口径 5.5、底径 4.1 厘米

敛口，弧形深腹，托盘边沿宽大，高圈足。内外满施黑釉，造型优雅。

宋 柳叶形铜匙

长 14.6 厘米

匙面呈柳叶形，微凹。匙柄细长，尾部向下弯曲。铜质保存较好，表面金属光泽依然莹亮。

宋代由于饮茶方式发生变化，茶匙在原来的量取作用的基础上又增加了击拂茶汤的作用。宋蔡襄《茶录》云："茶匙要重，拂击有力，黄金为上，人间以银铁为之。"梅尧臣有一首《次韵和永叔尝新茶杂言》，最后两句写道："石瓶煎汤银梗打，粟粒铺面人惊嗟。诗肠久饥不禁力，一啜入腹鸣咿哇。"其中的"银梗"就是指银茶匙。

宋 扁形铜匙

长 16.2 厘米
陈钢捐赠

匙面呈扇形，微凹，匙面背部中间有一出筋。匙柄细长、有弧度，捶打成扁平状，距柄端三分之一处饰三道弦纹。

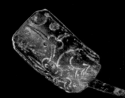

宋 刻云兔纹银匙

长 17.3 厘米

匙面呈委角长方形，錾刻仙兔口衔灵芝回首望月纹饰，匙柄细长平直，柄身錾刻弦纹。

宋 杏形铜匙

长 16 厘米

陈钢捐赠

匙面呈杏形，微凹。匙面宽的一侧连接匙柄，柄细长，略有弧度，中间为圆柱形，两端捶打成扁平状，既方便与匙面连接又利于提捏。

五

辽金茶事
Tea Culture of Liao and Jin Dynasties

游牧民族在北方建立的辽、金政权与两宋处于对峙、融合状态。受宋文化影响，这些游牧民族汉化程度较高，上层贵族亦有饮茶的习惯，饮茶方式、茶具形制和宋朝大同小异，也以汤瓶、茶盏、茶筅、茶匙、茶磨为主。在出土的辽金壁画中，不乏反映茶文化风貌的主题。

The Liao and Jin regimes established by nomadic peoples in the north confronted or fused with the Song Dynasty. They were highly influenced by the Song culture, and the upper class nobles also had the habit of drinking tea. Their way of consuming tea was almost the same as that of the Song. Tea bottles, cups, whisks, spoons and grinders, having substantial similarities to items prevalent in the Song, were used among the nomadic peoples, which were demonstrated in many murals of Liao and Jin Dynasties unearthed.

图 2.45　河北宣化辽张匡正墓壁画中的碾茶、煮茶场景
Scene of grinding and boiling tea showed in the Zhang Kuangzheng tomb of Liao Dynasty in Xuanhua, Hebei

图 2.46　河北宣化辽张匡正墓壁画"童嬉图"
Mural "Children Playing" from Zhang Kuangzheng tomb of Liao Dynasty in Xuanhua, Hebei

图 2.47 河北宣化辽张恭诱墓壁画"备茶图"
Mural "Tea Preparation" from Zhang Gongyou tomb of Liao Dynasty in Xuanhua, Hebei

图 2.48 北京石景山金代壁画中的点茶场景
Scene of *diancha* in the Jin Dynasty mural in Shijingshan, Beijing

图2.49 山西汾阳东龙观金代家族墓地壁画"茶酒位"及其局部图
Mural "Chajiuwei" and its detail of a family graveyard in Donglongguan Village, Jin Dynasty, Fenyang, Shanxi

金 黑釉油滴盏

高 6.8、口径 10.5、底径 5.1 厘米

敛口，斜弧腹，圈足。内外施黑釉，外壁施釉至半，有挂釉现象。釉面布满黄褐色沙粒状突起，内壁有红褐色和银灰色金属光泽的斑点，形似油滴。露胎处呈土黄色，胎质粗松。

金 耀州窑刻花水纹盏

高 5.6、口径 11.8、底径 4 厘米

圆唇,侈口,深腹,腹壁稍曲外鼓,矮圈足。盏心用尖状工具划出花纹,然后用篦具在花纹四周划出旋转流水纹。胎薄质坚,釉面光洁匀净,釉色青绿,微泛黄,秀气淡雅。

金 霍州窑白釉模印鸭纹盏

高 4、口径 10.8、底径 2.9 厘米

敞口，弧腹，矮圈足。盏内模印水波游鸭纹，内底有涩圈，便于叠烧。

霍州窑是金元时期山西地区的一处重要窑口，又名陈村窑、霍窑、西窑、彭窑等。明人曹昭《格古要论》记载："霍器出山西平阳府霍州……元朝戗金匠彭均宝效古定器，制折腰样者甚整齐，故名曰彭窑。土脉细白者，与定器相似，唯欠滋润，极脆，不甚值钱，卖古董者称为新定器，好事者以重价购之。"霍州窑瓷器产品主要有白釉、黑釉等，以白釉最具特色，其装饰手法有印花、刻花、划花、酱褐彩画花等，器形以碗、盘、罐、高足杯为主。

六

再传日本
Re-Spread to Japan

宋代，中日交流越加密切。中日间贸易的便利使得大量茶叶与茶具进入日本，日本吃茶的风气更盛，饮茶被看作是一种高雅风尚。南宋时期，入宋求法的日本僧人甚多。入宋僧带回南宋的禅宗，也带回精致繁复的茶会仪式和禅茶合一的茶道文化。

In the Song, China and Japan interacted more extensively. The Sino-Japanese trade facilitated the entry of more tea and tea utensils into Japan. Regarded as an elegant fashion, tea drinking became more popular. During the Southern Song Dynasty, many Japanese monks came to China. When they returned to Japan, they brought back with them the Zen Buddhism, complicated tea ceremony and culture of Zen-tea integration of the Southern Song Dynasty.

图2.50 荣西像
Portrait of Eisai

1. 荣西与《吃茶养生记》
Eisai and *An Account of Drinking Tea for Nourishing Life*

南宋乾道四年（1168年）和淳熙十四年（1187年），日本高僧荣西两度来我国学习佛经。他将中国的茶籽和饮茶法带回日本，并著《吃茶养生记》一书。荣西及其《吃茶养生记》在日本茶文化史上具有非常重要的意义。

Eisai, a Japanese eminent monk, traveled to China twice to study Buddhist sutras, first in 1168 and again in 1187 in the Southern Song Dynasty. He brought back to Japan both Chinese tea seeds and tea drinking methods. He also wrote a book titled *An Account of Drinking Tea for Nourishing Life*. Eisai and this text are of great significance in the history of Japanese tea culture.

图2.51 荣西禅师碑
Stele of Eisai

2. 径山茶宴传东瀛
Jingshan Temple's Tea Banquet Comes to Japan

径山坐落于今浙江余杭，历代多产佳茗，相传唐代法钦禅师曾"手植茶树数株，采以供佛，逾手蔓延山谷，其味鲜芳特异"。后世僧人常以本寺香茗待客，久而久之，便形成一套行茶的礼仪，后人称之为"茶宴"。日本僧人圆尔和南浦绍明先后于南宋端平二年（1235年）和开庆元年（1259年）至径山留学。回国时，他们带去茶种以及供佛待客的饮茶仪式，在日本广为传播。

Jingshan Temple, located in today's Yuhang, Zhejiang Province, was prolific with good tea in past dynasties. It is said that Tang master Faqin once "planted tea trees, picked tea leaves to worship Buddha. Later the trees spread over the valley, and the tea tastes fresh and fragrant". After that monks often treated guests with the tea produced in the temple, and a set of tea etiquette was formed over time, which was called "tea banquet" by later generations. Two other Japanese monks, Enn Ni and Nanpo Zyoumyou, came to Jingshan Temple to study in 1235 and 1259. When returning home, they brought tea seeds and tea drinking ceremony to worship Buddha and entertain guests, which spread widely in Japan.

图 2.52 日本静冈茶园中的圆尔诞生地之碑
Stele of the birthplace of Enn Ni in Shizuoka tea garden, Japan

图 2.53 浙江余杭径山寺
Jingshan Temple in Yuhang, Zhejiang Province

七

承前启后
Tea Culture in Transitional Period

元代国祚不长，在茶文化发展史上属于上承宋、下启明的过渡时期。元政府除了承袭宋代的北苑御茶园，另在福建武夷山九曲溪之第四曲溪畔设立御茶园，武夷茶开始成为贡茶。宫廷所用仍以团饼茶（即"蜡茶"）为主，民间一般饮叶茶和末茶。点茶法仍盛行于元代，相关茶具也延续宋代的风格，但也出现了用沸水直接冲泡散茶的饮用方式。

The Yuan Dynasty was short-lived. Its tea culture connected the preceding Song and the following Ming. Besides the Beiyuan imperial tea garden inherited from the previous dynasty, the Yuan government also set up another garden by the fourth stream of Jiuqu River in Wuyi Mountain, Fujian Province. Wuyi tea began to become tribute tea. Cake tea was still welcomed by the court, while loose tea and tea powder were generally drunk among the people. *Diancha* was still popular and the related tea sets were much like those used in the Song. Yet brewing loose tea with boiling water, a different way of making tea had appeared.

图2.54 内蒙古赤峰市元宝山元墓壁画"进茶图"
Mural "Tea Serving" of the tomb of the Yuan Dynasty in Yuanbaoshan, Chifeng, Inner Mongolia

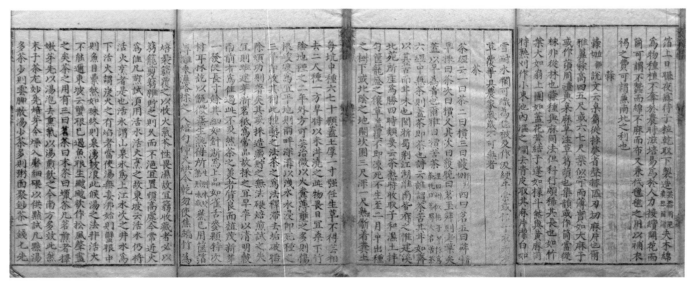

图2.55 王祯《农书》
Agricultural Book by Wang Zhen

王祯《农书》卷十"百谷谱十"之"茶"部分，介绍了茶树的栽培，以及茶的采造藏贮、碾焙煎试等，简短而精要。
In the "Tea" part of Volume 10 of *Agricultural Book* by Wang Zhen, the cultivation of tea trees, as well as the picking, storage, grinding, roasting and frying of tea is introduced briefly and concisely.

图 2.56 山西洪洞广胜寺元代壁画中的煮茶场景
Scene of boiling tea in the Yuan Dynasty mural of Guangsheng Temple in Hongdong, Shanxi

元 枢府釉印花折腰碗

高 5、口径 12、底径 4.3 厘米

侈口，斜腹，折腰，小圈足。白胎细腻，碗内外施卵白釉，釉层较厚，有乳浊感。内有印花装饰，并印有"枢府"两字。

圈足内无釉，因修足而留下鸡心点，这也是元代枢府釉瓷鉴别的特征之一。

元 钧窑直腹盏

高 5、口径 11、底径 4.5 厘米

敛口，鼓腹，圈足。砖红色胎，施青色乳浊釉，釉施至盏内壁一半处，外壁釉不及底，釉面凝厚。

元 龙泉窑青釉刻花云鹤纹盏托

高 4.5、口径 7、底径 5.5 厘米

呈圆盘形,盘口微微上翻,中空以承盏,斜弧腹,矮圈足。灰胎,盘面刻划一对飞鹤和云纹,纹饰简洁生动。器内外施青釉,釉色滋润肥腴。

第五章

茶韵隽永
—— 明代茶文化

Profound Tea Culture
Tea Culture of the Ming Dynasty

明代是古代茶文化创新变革的重要时代，在茶叶栽种、加工技艺、品饮方式、茶器茶具、名茶名品及茶道理念等方面，都对后世产生了深刻的影响。化繁为简的散茶开创了茗饮艺术的新局面，使茶更贴近人们的日常生活，也促成了茶具的推陈出新。此外，明代也是茶书著述最丰富的时期，延续了唐以来为茶著书的优良传统。

The Ming Dynasty is an important era in which innovation and transformation of ancient tea culture took place. It had a profound impact on later generations in tea planting, processing, drinking methods, utensils, notions of tea ceremony etc. The simple loose tea started a new chapter in the art of tea drinking, getting closer to people's daily life. It also generated innovation in the realm of tea utensils. In addition, the fine tradition of compiling texts for tea was carried down from the Tang to Ming, which was the period that boasted the most tea books.

一

废团改散
Abolishing of Cake Tea and Rise of Loose Tea

宋代进奉皇室的团饼茶虽然精美，但制法烦琐，费时费钱，到了元代已日渐衰微，而唐宋时期就已出现的散茶愈来愈受到人们的青睐。散茶蔚为风尚，是在明洪武二十四年（1391年）明太祖朱元璋下诏罢造龙团，改贡芽茶（散茶）之后。团茶和散茶此消彼长，不仅引发了饮茶方式和器具的变革，而且推动了散茶尤其是炒青绿茶的全面发展。明代的散茶有些早已湮没在时间的长河中，也有一些历久弥新，至今仍然为人啜饮。

The cake tea presented to the royal family in the Song Dynasty was exquisite, but its production was complicated,

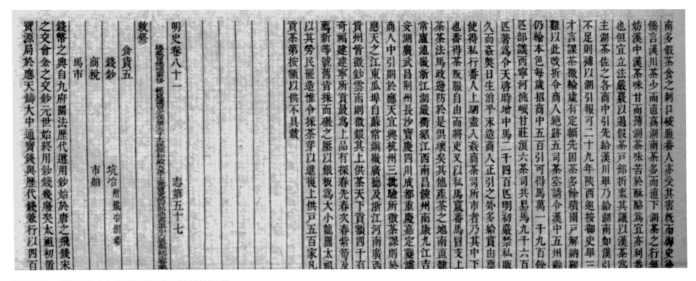

图 2.57 《明史》中朱元璋下诏改贡散茶的记载
Zhu Yuanzhang's edict collecting loose tea as a tribute item in *the History of Ming Dynasty*

time-consuming and expensive. In the Yuan Dynasty, it gradually declined and the loose tea that emerged in the Tang and Song was increasingly favored by people. In 1391, the first Ming Emperor Zhu Yuanzhang issued an edict stopping the production of dragon-phoenix cake tea and demanding bud tea (loose tea) as tribute instead. After that loose tea took hold. The fall of cake tea and rise of loose tea not only triggered the reform of tea drinking methods and utensils, but also promoted the development of loose tea, especially roasted green tea. Some loose teas of the Ming Dynasty have long been lost, while some are still being drunk today.

二

瀹饮慢品
Brewing and Sipping Tea

图 2.58 朱元璋像
Portrait of Zhu Yuanzhang

随着明初罢贡龙凤团茶，茶叶的品饮也一改唐宋时期的烦琐，趋于简化。"今人惟取初萌之精者，汲泉置鼎，一瀹便啜，遂开千古茗饮之宗。"用沸水冲泡叶茶，"旋瀹旋啜"的瀹饮法，有崇尚自然、返璞归真之妙。

由于茶叶不再碾末冲点，宋代流行的茶臼、茶碾、茶磨、茶罗、茶筅及黑釉盏皆废弃不用，茶壶和茶杯的组合代之而起。景德镇的白瓷及在此基础上发展起来的青花、五彩、斗彩和颜色釉瓷茶具成为明人的选择。紫砂茶具在明代中期开始出现，逐渐成为重要的泡茶器，并一直流行至今。

As the dragon-phoenix cake tea was stopped being collected as tribute in the early Ming, tea was prepared in a simpler and easier way than in the previous Tang and Song Dynasties. As described by a scholar, "Today, people only take the fine tea buds, pour spring water into the pot, brew tea with boiling water and then sip the tea. Such a method sets an example for the future way of making tea." This method call "yue-yin" for the preparation of the beverage brought back the original simplicity of tea.

As tea leaves were no longer ground into powder and whisked, tea mortars, grinders, sieves, whisks and black-glazed cups popular in the Song Dynasty were abandoned, and the combination of teapot and teacup replaced them. The white porcelain of Jingdezhen and the blue-and-white, multicolored, clashing color and color-glazed porcelain tea sets became the choice of the Ming people. Purple clay tea utensils came on the stage in the mid-Ming, and gradually became important tea makers, which are prevalent even now.

三

茶书撰著
Tea Books

图2.59 明 文徵明《品茶图》(局部)
Painting of Tea Tasting (Partial), Wen Zhengming, Ming Dynasty

自唐代陆羽为茶著书以来，历代都有茶书之出。明代，尤其是明代中后期，茶书纷纷涌现，今人能看到的有五十几种，约占中国古代茶书的一半。明代茶书不乏佳作，如朱权撰《茶谱》、陈继儒撰《茶董补》，于清饮有独到见解；田艺蘅撰《煮泉小品》、陆树声撰《茶寮记》，反映高士情趣；许次纾撰《茶疏》，精于茗理。

After the Tang, when the first treatise on tea was composed by Lu Yu, tea books constantly emerged. In the Ming Dynasty, especially in the middle and late periods, tea books came out one after another. The surviving tea-focused texts are more than 50, accounting for about half of the ancient Chinese works on tea, among which some texts are worth mentioning. For example, *Tea Manual* (Zhu Quan) and *Supplement to Tea Records* (Chen Jiru) hold unique views on tea drinking. *Essays on Tea & Water* (Tian Yiheng) and *Records of Tea House* (Lu Shusheng) reflect the taste of literati. *Annotation to Tea* (Xu Cishu) is sharp at tea culture.

图2.60 明 丁云鹏《煮茶图》（局部）
Painting of Boiling Tea (Partial), Ding Yunpeng, Ming Dynasty

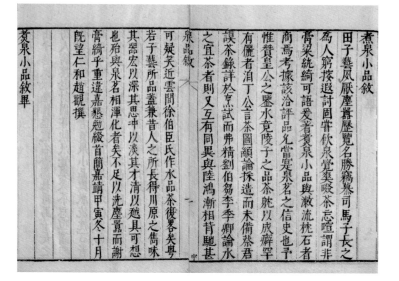

图2.61 陆树声《茶寮记》
Records of Tea House by Lu Shusheng

图2.62 田艺蘅《煮泉小品》
Essays on Tea & Water by Tian Yiheng

图2.63 屠本畯《茗笈》
The Record of Tea by Tu Benjun

明万历 青花折枝花纹提梁壶

通高 19.5、口径 6.8、通宽 18.8、底径 8.7 厘米

矮直口，短颈，弧肩，圆鼓形腹，矮圈足，肩部两侧向上起虹式提梁，腹部一侧置弯流，配宝珠纽盖。整器装饰青花折枝纹，青花发色淡雅，釉面滋润。

明代，饮茶史上出现了一个重大变革，延续几代的团饼茶被散茶所代替，散茶瀹泡的饮用方式开始渐渐取代"唐煮宋点"，真正意义上的茶壶诞生了。

通高 14、口径 5、底径 6.5 厘米

　　直颈，溜肩，腹部渐收，腹部一侧置细长流，另一侧置环形把，配宝珠纽盖。壶身两面开光内绘菊花纹饰，民窑青花绘得较为草率奔放。

　　菊花纹是瓷器装饰纹样之一，在明代早中期较常见，明晚期出现较少。

明 "用卿" 款紫砂金钱如意壶

高 29.8、口径 13.8、底径 13.6 厘米

唇口，宽颈，鼓腹，内凹底，弯流，曲把，器形硕大。盔式盖贴塑如意纹，上立镂雕金钱纹纽。壶身一侧刻"瓦瓶新汲三泉水，纱帽笼头可自煎"诗句，落款为草书"丁卯年，用卿"。制作年代应为 1627 年。

陈用卿，明代天启、崇祯年间人。《阳羡名陶录》称其书刻"落墨拙而用刀工"。明代名士张岱在其《陶庵梦忆》中称："宜兴罐，以供春为上，时大彬次之，陈用卿又次之。"可见此壶作者当年与时大彬齐名而稍逊一筹，不失为一代制壶名师。陈用卿壶存世量很少，足见此壶之珍贵。

高 16、口径 9.5、底径 12.8 厘米

宽唇，直口，矮颈，垂腹，圆底内凹。肩部一侧置炮管流，另一侧置扁圆形把，配大圆纽盔式盖。壶底钤印"花晨月夕 舍此不可"八字双行楷书款。花晨月夕特指美好的时光和景物。清代汪汝谦的《画舫约》有"花晨月夕，如乘彩云而登碧落"之句。王复礼的《茶说》则曰："花晨月夕，贤主嘉宾，纵谈古今，品茶次第，天壤间更有何乐？"可见，花晨月夕，以紫砂壶来啜饮茗茶，乃天下一大乐事也。

整壶采用明代流行的砖红泥制作，壶身满布细细的桂花砂，珠粒隐隐，更自夺目。整器观之，如一披红色袈裟的老衲，盘跏趺坐于蒲团之上，淡定稳重。

高 16.5、口径 7、底径 8 厘米

唇口,短颈,丰肩,腹部内收。通体施蓝釉,釉色莹润。

明 青花"上品香茶"罐

通高 6.8、口径 3.2、底径 4 厘米

直口，短颈，丰肩，斜腹，配平顶盖。罐外壁绘青花双马，马背上各立一块方形招牌，一牌上书"上品"，另一牌上书"香茶"。罐盖上绘七瓣花卉一朵。此盖罐为明代市井茶叶铺中装茶叶的器皿。

明 紫砂绞泥六方茶叶罐

高 8.5、口径 2.7 厘米

直口，短颈，折肩，腹部为六方体，腹下渐收，平底。器肩部、腹部均采用绞泥工艺，饰以不规则条状白泥，如行云流水，颇为雅致。

明末清初 紫砂六方茶叶罐

通高 18.6、口径 8.8、底径 14.8 厘米

直口，折肩，六方体腹到底部渐收，带盖。上腹部六个转折角处装饰大如意纹，给端庄大气的罐体增添了几分妩媚之感。平底装饰三个磬形小矮足，使罐体更加挺拔俊朗。泥料带些淡墨色，且掺了细细的黄金砂。《阳羡茗壶系》里提到的"细土淡墨色"说的就是这种泥料。茶叶罐容量大，应该是盛放平时不经常取用的茶叶，密封性能极好。

明 剔犀如意云纹漆盏托

高 9、口径 9、底径 8 厘米

由圆形盏、葵瓣式盘、高足组成。通体剔犀工艺，雕如意云头纹饰，花纹刀口侧面露出黑漆线。托内髹黑漆，乌黑黝亮，有自然断纹。

剔犀又称云雕，其工艺先以两色或三色漆相间漆于胎骨上，每一色漆都由若干道漆髹成，至相当的厚度后，斜剔出云钩、回纹等图案花纹，故在刀口断面可见不同的色层。

第六章 茶意不歇
——清代茶文化

Endless Charm of Tea
Tea Culture of the Qing Dynasty

明代萌芽的资本主义经济在清代有了进一步发展。茶业方面，清代茶树栽培、茶叶加工技术更为完善，茶区面积扩大，产量提高。名茶纷纷涌现，品饮方式更趋多样。城乡茶馆、茶庄林立。无论在宫廷还是市井，茶都大受欢迎。此外，中国的茶叶开始大规模地出口欧洲。

The capitalist economy sprouted in the Ming further developed in the Qing period, in which tea cultivation and processing techniques were improved, tea-growing areas were expanded and the tea output was increased. Famous teas emerged, and the drinking methods became more diversified. Many tea houses or shops were built in cities and villages. Tea was very popular both in the court and among the masses. In addition, Chinese tea began to be exported in large quantity to Europe.

一

华茶出口
Export of Chinese Tea

中国的茶业曾经领先于世界各国。华茶出口最早可以追溯至17世纪，从那时直至19世纪初，中国作为世界上唯一的茶叶出口国，二百年来独步国际茶叶市场。1838年首批印度茶出现于英国伦敦市场时，这种局面才被打破。此后，华茶的海外市场份额逐渐被印度茶、锡兰茶和日本茶蚕食。

Chinese tea industry once ranked the first in the world. The export of Chinese tea can be traced back to the 17th century. China, as the only tea exporter in the world, dominated the international tea market for 200 years from the 17th century to the early 19th century. The situation changed only when Indian tea first appeared in London market in 1838. Since then, the overseas market share of Chinese tea was gradually eroded by teas from India, Ceylon and Japan.

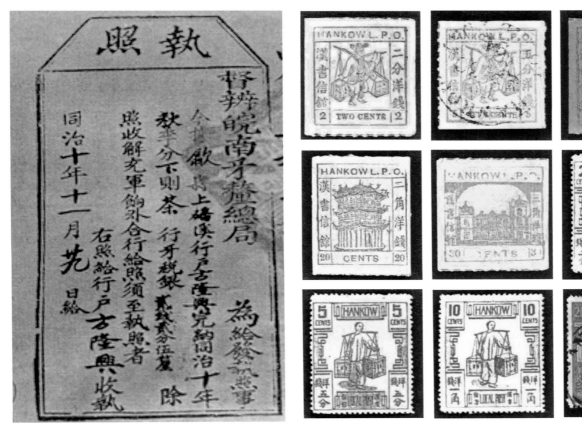

图2.64 箱茶运销茶引
Tea trade certification for box tea transportation and selling

图2.65 茶担图商埠邮票
Stamps with portrait of a tea carrier in a port

图2.66 清代茶叶装箱外运
Packing tea for transportation, Qing Dynasty

图2.67 外销茶叶装箱
Packing tea for export

● 哥德堡号 The Gotheborg

"哥德堡号"是大航海时代瑞典东印度公司著名的远洋商船,曾三次远航至中国广州。1745年1月11日,"哥德堡号"从广州启程回国,船上装载着大约700吨的中国商品,包括茶叶、瓷器、丝绸和藤器等,其中茶叶2677箱。经过8个月的航行,在离哥德堡港大约900米的海面,"哥德堡号"突然触礁沉没。20世纪80年代,瑞典人发现了沉睡海底的"哥德堡号"并对之进行水下考古,打捞出大量的瓷器和茶叶。1995年,瑞典新东印度公司着手复原"哥德堡号",通过十年努力,终于打造完成这艘以18世纪工艺制作的新"哥德堡号"。2005年10月2日,"哥德堡号"扬帆出海,重走260年前的航路。2006年7月18日,"哥德堡号"抵达具有历史意义的终点——中国广州,然后驶往上海。"哥德堡号"在中国逗留期间,中国的多个茶叶产地向其赠送了当地名茶,希望"哥德堡号"能再次将中国茶文化带往欧洲。

The "Gotheborg" was a renowned ocean-going merchant ship of the Swedish East India Company during the Age of Discovery. It made three voyages to Guangzhou, China. On January 11, 1745, carrying approximately 700 tons of Chinese goods, including tea, porcelain, silk, and wickerware, the "Gotheborg" headed back to Sweden from Guangzhou. Eight months later, only about 900 meters away from the port of Gotheborg, the vessel suddenly struck a rock and sank. In the 1980s, the wrecked "Gotheborg" was discovered by the Swedish people, thus underwater archaeology was conducted and a large number of porcelain and tea were salvaged. In 1995, the Swedish East India Company initiated the reconstruction of the "Gotheborg". They spent ten years creating a replica exactly made with the same techniques back in the 18th century. On October 2, 2005, the new "Gotheborg" set sail on the same journey 260 years ago. A historic moment was on July 18, 2006, when the ship reached Guangzhou, and then Shanghai. During its stay in China, the "Gotheborg" received various best-quality teas from different regions. The Chinese people wish this brand-new "Gotheborg" could once again spread the Chinese tea culture to Europe.

清 瑞典"哥德堡号"沉船茶样

茶样呈黄褐色,带梗结块,基本已氧化变质,难以辨认茶叶品种。

"哥德堡号"是瑞典东印度公司的重要商船。1731—1806年的75年间,瑞典东印度公司共派出37艘商船,完成了从瑞典哥德堡到中国广东的132次远洋航行,这些商船主要从事茶叶、丝绸、瓷器、香料贸易。1743年,"哥德堡号"开始了第三次到中国广东的航行,于1745年9月12日在快要驶进哥德堡港口时沉没,船上载有各种中国茶叶约370吨。1986年,瑞典开始了水下考古,除了打捞出大量的瓷器外,还有大量茶叶。这些茶叶盛放在特制茶箱内,又有锡罐严密封装,沉没海底两百多年还没有完全腐蚀。透过这些茶叶,我们看到的是华茶外传的足迹;根据这些茶叶,我们可以追溯中国茶叶输入欧洲以及欧洲饮茶之风兴起的历史。

清 黑漆描金人物纹茶叶盒

高11、长23、宽14.5厘米

呈不规则八方形,盖面及器身以黑漆描金绘制合家欢场景。打开盒盖,内有一对锡茶罐,锡罐盖及肩部均刻有花卉纹。

该茶叶盒系当时的外销茶具,其黑漆描金工艺为当时最受欧洲欢迎的工艺之一。内置的两个茶叶罐一般会放置红茶和绿茶两种不同的茶叶,有时配合放置一个糖罐或者一个用以混合茶叶的容器。茶叶盒配有锁孔,可以上锁,从侧面证明当时欧洲茶叶价格昂贵。随着茶叶价格的大幅度下降,这类带锁的茶叶盒也逐渐不再使用。

晚清 "合兴祥"宝号箱茶运单

纵 23、横 19 厘米

这是清代晚期的一张茶叶运输清单，落款为江耀华。江耀华系晚清徽商中非常有代表性的茶商，曾留下大量与茶叶经营有关的文书档案。

据《近代徽州茶业兴衰录》一书梳理芳坑江氏家族业茶经历，可知江耀华为茶商江有科之孙，祖孙三代均以经营外销茶为主要项目，在歙县开设茶号，就地采购茶叶，经加工制作后，运往广州外销至西方。当时徽茶运至广州的路线主要是经由江西赣江溯流而上，越大庾岭而达广州。在交通不便、关卡林立的条件下，迢迢千里运茶入粤十分艰难。

清道光年间，江有科父子经营茶叶贸易获利颇厚，曾在芳坑大兴宅第。但至清咸丰元年（1851年），太平军起义，江氏父子贩茶入粤的道路被阻，茶叶生意大受影响。家道中落之后，江耀华便在一家茶号中当佣工，之后通过李鸿章的介绍，负责谦顺安茶栈的事务。有了一定的积累之后，江耀华开办了自己的茶号。当时徽州的茶号都是临时性的商业机构，名称和地址并不固定。据现存江耀华的账册、札记、信函，可知他的茶号曾用过的字号招牌有"永盛怡记""张鼎盛""合兴祥""泰兴祥"等，这些茶号大都设在屯溪。当时的屯溪是徽州茶叶的集散地，茶号设在这里既便于从徽州各地收购毛茶，又便于把加工后的箱茶运销外地。

这张运单便是清代晚期江耀华经营的"合兴祥"茶号的运输单据，这张运单上注明了运送茶叶的箱数和重量，用于通过当时杭州拱宸桥关卡。这一时期徽州的茶叶多通过上海出口，而当时徽州至上海的路线，是从屯溪搭船沿新安江东下到杭州，在杭州过塘以后再抵达上海，拱宸桥是当时杭州重要的口岸，对各种物资进行卡关检查。

合興祥寶號箱茶叁壹七件計重壹萬柒千○十三觔

好正税○三亭
嘸位觔○陸
叩用○津貼銀三餘
保險元埔
此申莊元 決番廿貳九五六半
以申莊元 五拾三貳五五
奴申莊元 貳零九斗
呅申莊元 拾貳半
申莊元 三貳生卜
申馬駝搢之✓

共莊元 叁百拾貳貳玖斗

洪會卿先生臺 午月 兄 慎昌湯華

二 茶馆风情
Various Tea Houses

清代，茶馆遍布城乡各地，社会功能也有拓展，出现了为不同人群服务的特色茶馆，如专供商人洽谈生意的清茶馆，表演曲艺说唱的书茶馆，供文人笔会、游人赏景的野茶馆，供茶客下棋的棋茶馆等。清代是继宋之后中国茶馆文化的第二个兴盛期。

In the Qing Dynasty, tea houses were distributed all over the urban and rural areas. Their social functions were also expanded. Special tea houses serving different groups of people appeared. A tea house could be a place where businessmen met to work out deals; performers gave their performances; literati and tourists enjoyed the scenery; tea drinkers played chess. The Qing Dynasty was the second prosperous period of Chinese tea house culture after the Song.

三 清饮成趣
Tea Drinking

清代饮茶方式沿袭明代，以茶叶冲泡清饮为主。相较于前代，清代茶具的材质更加丰富，釉色与品种也更多样化，除青花和各种颜色釉外，又增加了粉彩、珐琅彩等新品种。清代茶具主要有盖碗、茶杯、茶船、茶盘、茶叶罐、茶壶等。

Tea in the Qing was prepared by brewing with hot water, a method passed down from the previous dynasty. Compared with the Ming, tea utensils in the Qing were more diversified in materials, glaze colors and varieties. In addition to blue-and-white and various colored glazes, famille rose and enamel painted porcelain were created. Tea wares frequently used in the Qing were covered bowls, cups, tea boats, tea trays, tea caddies, tea pots, etc.

● 盖碗 Covered Bowls

一般由盖、碗、托三部分组成，象征着"天、地、人"三才，也有些盖碗仅有盖和碗两部分。作为茶具的盖碗在明末清初时开始出现，后成为清代主要的茶器。

It generally consists of a cover, a bowl and a saucer, symbolizing "heaven, earth and man". Some covered bowls only have a cover and a bowl. Covered bowls began to appear in the late Ming and early Qing Dynasties, and later became the main tea wares in the Qing Dynasty.

● 盏托 Saucers

汉代已有盏托，唐宋时期颇为流行。明代，盏托演变为舟形，由此得名"茶舟"或"茶船"。清代茶船蔚为风尚，形制各异，有舟形、元宝形、海棠花形、"十"字花形等，材质则有陶瓷、漆木、锡银等，异彩纷呈。

The saucer could date back to the Han Dynasty, and gained great popularity in the Tang and Song. In the Ming Dynasty, the saucer began to take the shape of a boat, hence the name of "tea boat". In the Qing, tea boats were shaped into such forms as boat, gold or silver ingot, begonia flower, cross, etc., and materials used are ceramic, lacquer, wood, tin, silver and so on.

● 紫砂壶 Purple Clay Teapots

经过明代的初步繁荣，到了清代，紫砂茶具迎来新的创作高峰。尤其是清嘉庆、道光之后，文人雅士相继加入制壶行列，大大提高了紫砂茶具的人文内涵。他们以紫砂为载体，发挥其诗、书、画、印的才情，为后人留下了不少精美的紫砂艺术品。

Purple clay tea wares developed in the Ming Dynasty. By the Qing, it ascended to the peak of creation. Notably, after periods of Jiaqing and Daoguang, literati joined in the making of pots, which greatly improved the humanistic connotation of purple clay tea wares. They expressed themselves on purple clay tea wares with poems, calligraphy, painting and seals, creating and leaving many fine purple clay artworks for later generations.

四 茶庄茶号
Tea Shops and Firms

随着国内外茶叶贸易的发展，清代大量出现买卖茶叶的茶庄、茶号、茶行、茶栈。其中茶庄、茶号以零售业为主，茶行是茶叶买卖双方的中介，茶栈主要从事出口茶叶的收购与加工。许多茶庄、茶号还经营特色名茶，如上海"汪裕泰"茶号以专售安徽的红茶、绿茶而闻名，杭州"翁隆盛"茶号则以出售春前、明前、雨前的西湖龙井茶（"三前摘翠"）而极负盛名。

A large number of tea shops, *hong* or firms were set up in the Qing Dynasty with the development of tea trade at home and abroad. Tea shops were mainly engaged in retail business, the tea *hong* was the intermediary between the buyers and sellers of tea, while tea firms chiefly purchased and processed exported tea. Many tea shops also sold famous teas. For example, Shanghai Wangyutai tea shops were famous for exclusively selling black tea and green tea from Anhui, and Hangzhou Wenglongsheng tea shops were well-known for selling West Lake Longjing tea produced around the Qingming Festival and before Grain Rain.

图2.68　清 吴昌硕《品茗图》
Painting of Tea Tasting, Wu Changshuo, Qing Dynasty

图2.69　清 任伯年《煮茶图》
Painting of Boiling Tea, Ren Bonian, Qing Dynasty

图 2.70 水彩画《广州茶叶铺》，约 1830 年
Guangzhou Tea Shop, watercolor, ca.1830

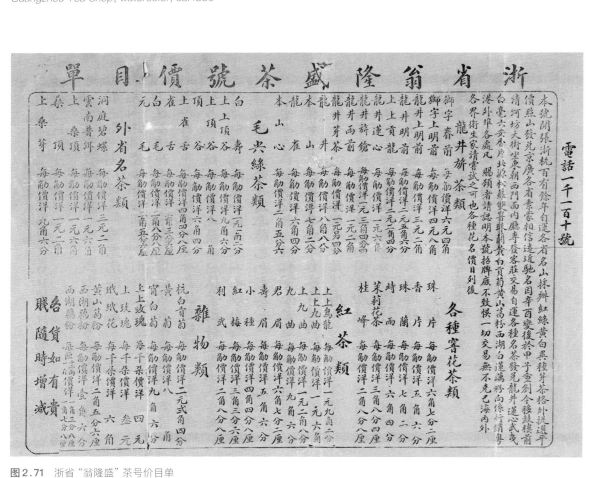

图 2.71 浙省"翁隆盛"茶号价目单
Price list of Hangzhou Wenlongsheng tea shop

清咸丰十一年 照验

纵 34、横 19.5 厘米

倪龙江捐赠

纸质。左下、右上与右下角均截角，上印黑色楷书，内容为"钦加知府衔护理江南徽州府正堂加十级纪录十次伊 给发引照事照得咸丰拾壹年徽郡出境引茶 部引未奉颁发今据 歙 邑茶商 新茂 配茶拾担给总照壹道照验截角放行俟引颁到按照截引缴销须至信照者 咸丰拾壹年拾月十八日给"。

清代，政府对徽州商人的茶叶运输加强了管理。徽州茶商贩运茶叶均须持有地方衙门所颁的"引票"（或"照验"），以备关卡查验。茶商若无引票而贩运茶叶，一经关卡查出，即遭严惩。茶叶运输管理的加强从一个侧面反映了茶叶贸易的兴盛。

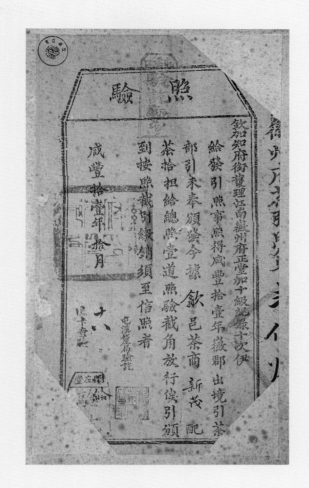

清同治十一年 售茶园户捐照

纵 23.5、横 15 厘米

倪龙江捐赠

纸质。为清同治十一年（1872年）四月十六日清政府颁发给售茶园户的凭证。茶叶作为商品，历史上一直是政府控制的经济作物之一，从唐代开始，茶叶税收一直是政府财政收入的重要组成部分。清代晚期，清政府对茶叶流通的控制更为严格，此即反映清政府对售茶园户在买卖茶叶过程中征收捐税的实物资料。

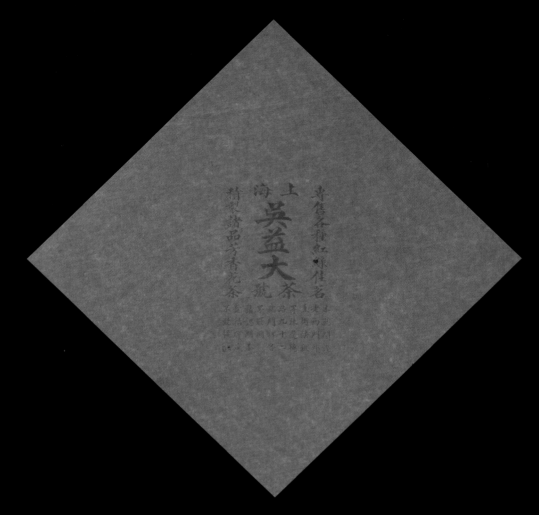

晚清 上海"吴益大"茶号包装纸

纵 25.5、横 27 厘米

纸质。上印绿色字样,系晚清上海"吴益大"茶号广告纸,上标明茶号经营地址及经营茶类。

晚清 "眼生芝珠" "眼生熙春" 茶叶包装纸

纵 21、横 30.5 厘米

两件。纸质。晚清外销茶叶包装纸，多张贴于茶箱外侧。这类包装纸一般会体现茶叶品种、茶号、产地等信息，既有标明产品型号的功能，又兼有广告的效果。

包装纸中心开光内写有"世间绝品谁能识，闲对《茶经》忆古人。味兰逸客"，左侧红印为"清香好品"，右侧红印为"金怡泰制"。四周圆形开光内写有"眼生芝珠"或"眼生熙春"，间以白描竹、菊、梅、兰图样。

"芝珠""熙春"均系晚清徽州地区生产的绿茶品种。清何润生撰《徽属茶务条陈·计开》："徽属产茶，以婺源为最。每年约销洋庄三万数千引，歙、休、黟次之。绩溪又次之。该四县每年共计约销洋庄四五万引，均系绿茶。……绿茶内分三总名，曰珠茶，曰雨前，曰熙春。熙春内分四等，曰眉正，曰眉熙，曰副熙，曰熙春；雨前分五等，曰珍眉，曰凤眉，曰娥眉，曰副娥，曰芽雨；珠茶内分五等，曰虾目，曰麻珠，曰珍珠，曰宝珠，曰芝珠。各名色中，以虾目、珍眉为品之最上，凤眉、麻珠次之，眉熙、珍珠、娥眉又次之，宝珠、副娥、副熙更次之，最下乘者，芝珠、芽雨、熙春三等。"

"眼生"是徽州地区外销茶名中常见的前缀，意为质量上佳。

晚清 "同茂东官礼茶食" 印板

长 14.3、宽 8.2、厚 2.4 厘米

此为晚清"同茂东"茶食店的广告包装印板,上有"官礼茶食""本铺开设在唐田街十字口东路北"字样。

茶食最早在《大金国志·婚姻》中载有:"婿纳币,皆先期拜门,亲属偕行,以酒馔往……次进蜜糕,人各一盘,曰茶食。"在中国的饮茶生活中,茶点是不可或缺的,特别是江南的茶食店品种丰富,琳琅满目。此印板就是晚清众多茶食店之一,用来印刷广告包装纸,招徕顾客。

清 粉彩鹊桥相会盖碗

通高 9、口径 10.4、底径 4.4 厘米

由碗盖和碗两部分组成。盖直口，弧面；碗花口，弧腹，矮圈足。盖外壁以粉彩绘"鹊桥相会"图案，碗外壁绘一周通景式七仙女与董永相会画面，粉彩描绘人物、山水、树石等，细腻生动。捉手内、圈足内以及碗内壁均施松石绿彩，捉手内底、圈足内底以矾红彩书"江正隆制"四字三行篆书款。

盖碗茶具盛行于清代，无论宫廷皇室、达官贵人及市井民间，皆重盖碗茶。各种材质、各种纹样的盖碗竞相出现，给中国的茶具大家族增添了不少光彩。

清嘉庆 青花粉彩人物纹盖碗

通高 7.1、口径 11.5、底径 3.6 厘米

由碗盖和碗两部分组成。碗敞口，斜深腹，圈足，盖小于碗，倒扣时形制与碗相同。碗口沿与盖口沿饰青花如意纹，碗圈足上部及近捉手处饰变形莲瓣纹，中间主体部位以粉彩绘庭院仕女，捉手内与圈足内书"大清嘉庆年制"六字三行篆书款。

清光绪 粉彩过枝瓜蝶纹盖碗

通高 8.1、口径 10.3、底径 3.9 厘米

由碗盖和碗两部分组成。碗口沿及盖口沿饰以金彩,器内外均施粉彩,装饰十分考究。在洁白细腻的白瓷上,用淡黄、浅绿、深绿、粉红、矾红、白色等色彩,渲染出绵绵瓜瓞的纹饰,瓜藤的枝蔓沿着竹竿一直延伸到盖面及碗内,以此来象征子子孙孙绵延不绝,繁荣昌盛。碗底和提手内以红彩书"大清光绪年制"六字双行楷书款,书写十分规整,出自御窑厂专门写款的工匠之手。

清 豆青地粉彩鱼藻纹盖碗

碗（含盖）：通高 9.6、口径 10.6、底径 3.8 厘米
托：高 2.6、口径 10.6、底径 4.2 厘米

　　由盖、碗及托三部分组成。盖面弧形，上有捉手便于提拿；碗撇口，弧腹，矮圈足；托敞口，弧腹，矮圈足，托内中心下凹以承盏。通体施豆青釉作地，盖、碗外壁及托内壁以矾红彩绘金鱼九尾，以深绿、浅绿绘藻类植物树枝，金鱼眼睛以黑彩点出，鱼体生动，游弋于藻类植物之中。

清 松石地粉彩花蝶纹茶船

高 3.8、长 18.8 厘米

呈船形，浅圈足，器底内凹。通体以松石绿釉为地，内壁以粉彩绘一组花卉、蝴蝶，花蝶绘画写实，栩栩如生。全器色彩鲜艳，绘画工整。

清道光 松石地凸白花"十"字花形茶船

高 3.1、长 14.7 厘米

呈"十"字花形，敞口，弧腹，矮圈足，内底心下凹以承茶杯。器内外满施松石绿釉，上以白彩凸绘缠枝花卉纹。圈足口沿露胎，呈灰白色。圈足内以矾红书"大清道光年制"六字三行篆书款。

清乾隆 仿雕漆茶船

高 4.5、长 14.1 厘米

　　呈船形。外壁采用仿雕漆工艺，在瓷器表面雕刻出锦地纹及连续回纹，上施矾红釉。器内施金彩，由于经常使用，金彩已剥落。

　　瓷仿雕漆工艺出现于清乾隆时期，在瓷胎上先印纹样，经雕剔，入窑素烧后，施矾红釉，低温二次烧成。乾隆朝的瓷器在工艺及品种上都可以说达到了瓷器发展史的顶峰，这种用瓷器仿木雕漆器的工艺令人叹为观止。朱琰《陶说》记，乾隆时制瓷业能仿"饫金""髹漆""竹木"等多种手工艺品。此茶船形、色、纹均具雕漆器的质感，乍看几可乱真。

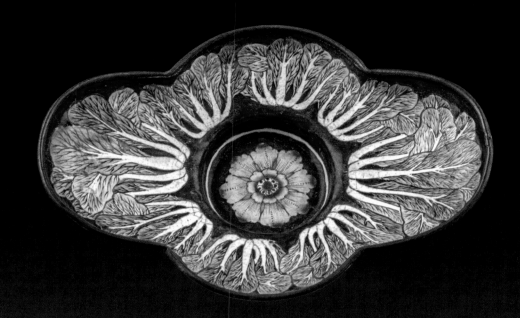

清 铜胎画珐琅白菜纹茶船

高 2.1、长 15.2 厘米

呈"十"字花形，内底心下凹以承茶盏。以铜为胎，通体以蓝色为地，内壁绘珐琅彩大白菜八株，描绘栩栩如生，内底绘宝相花一朵，用色十分淡雅。

铜胎画珐琅在清康熙时期由欧洲传入中国，画珐琅多以红铜为造型，表面以白釉为地，再用彩釉描绘图案，经焙烧、镀金而成。既体轻、明亮，又绚丽多彩，适合选取丰富的装饰题材。

清乾隆 青花釉里红四方茶叶瓶

高 18.6、口径 4.4、底长 15.8 厘米

呈四方体，圆直口，平肩，平底。口部涩胎无釉，肩部以青花釉里红饰折枝花。瓶体四面绘两组主题纹饰，其中两面绘山水人物纹，远山近水，山体披麻皴法描绘，山间有一茅舍，两高士策杖相遇，正在寻师问道，远处有两叶扁舟泛于湖上；另两面绘山石花卉纹，山石及树叶、树干用青花描绘，花卉则用釉里红描绘。茶叶瓶的主要作用是贮存茶叶，保持茶叶干燥。

清康熙 青花如意纹茶叶瓶

通高 19.2、口径 3.5、底径 5.6 厘米

方唇，直口，短颈，丰肩，弧腹，近底内收，圈足。口沿刮釉一周，外壁以青花装饰，肩部绘四朵如意云纹，云纹内以青花为地勾勒花朵纹，花朵内留白，胫部同样辅以四朵如意云纹，腹部间绘四朵折枝花纹。这类如意花卉纹系清康熙时期比较流行的式样，多见于外销器。修足规整，足端无釉。器盖为后配金属盖。

清"宜富当贵"紫砂方砖壶

高 7.4、通宽 10.3 厘米

造型呈四方砖形,壶体、壶纽、壶盖、壶把及壶流均设计成四方体状,因此俗称为"方砖壶"。壶腹一侧横印"宜富当贵"隶书体,摹自汉瓦当纹饰。

通高 19、口径 5.2、底径 9.9 厘米

以金属锡为内胎，锤打成形。宝珠纽盖，肩部置高提梁，椭圆形腹略下垂，壶腹一侧置弯形流，平底下接三足。壶肩腹部及盖面以椰壳作为装饰，肩部和胫部凸起两道双弦纹，腹部浅浮雕渔家乐纹饰，描绘渔夫们欢乐的劳动生活情景。劳动场景、生活场景和自然景色融合在一起，弥散着丰富的生活情趣。此提梁壶工艺复杂，制作精良，人物刻画细腻，雕刻技法娴熟。

早在公元 9 世纪以前，海南人就以椰壳制成器皿。明末清初学者屈大均在《广东新语》中记载，广东椰壳甚坚，可以"横破成碗，纵破成杯，以盛酒"。

清 竹黄六方茶壶桶

高 15.9、口长 14、底长 14 厘米

呈六方体。桶身材质为竹黄，口、底及盖沿为竹片髹黑漆。桶盖面镶嵌一对桥形提梁。六方壶身薄意浅刻竹石、鱼藻及梅兰纹饰。茶壶桶正面的中间口沿开一四方小口，系壶嘴安放之处，小口四周镶嵌如意形铜合页。

茶壶桶系保温茶壶之用，在热水瓶出现前，古人用茶壶泡茶，把壶放入其中，再在边缘填满丝棉絮，具有保温效果。此茶壶桶制作讲究，系中等以上人家使用，盖内印有"浙宁、江南、五汉郭顺记"字样，是制作该茶壶桶的商号或坊号。

翻黄工艺是中国民间工艺之一，把楠竹锯成竹筒，去节去青，留下薄层的竹黄，经过煮、晒、压平后，胶合或镶嵌在木胎上，然后磨光，再在上面刻绘人物、山水、花鸟等纹样，色泽光润，类似象牙。

五

宫廷茶事
Tea in the Royal Court

清代康、雍、乾三朝皇帝，一朝比一朝爱茶。清代的贡茶省份由明代的福建、浙江、南直隶、江西和湖广五省扩展到全国的十三省产茶区，贡茶的数量与品种也远超前代。清代宫廷的饮茶文化也对民间饮茶产生了一定的影响，形成了清代独特的茶文化。

Qing emperors Kangxi, Yongzheng and Qianlong loved tea very much. The provinces paying tribute tea in the Qing expanded from Fujian, Zhejiang, Nanzhili, Jiangxi and Huguang in the previous dynasty to 13 provinces nationwide in this period. The amount and type of tribute tea far exceeded those in the Ming. The imperial house's tea culture exerted certain impact on the common people, leading to a unique tea culture in the Qing.

1. 茶宴 Tea Feast

乾隆首倡重华宫茶宴，每年于正月择吉日举行。据记载，清代于重华宫举行的茶宴多达数十次，以赋诗联句和饮茶为主要内容。

Emperor Qianlong initiated the Chonghua Palace tea feast, which was held every year on the auspicious day of the first lunar month. Records have shown that dozens of tea feasts were held in the palace in the Qing Dynasty, focusing on poetry composing and tea drinking.

图 2.72　清 姚文瀚《紫光阁赐宴图卷》中的茶宴
Tea feast in the *Scroll of Ziguang Pavilion Feast*, Yao Wenhan, Qing Dynasty

2. 御茶房 Imperial Tea House

乾清宫东庑最北的三楹房间为"御茶房",匾额为康熙手书。御茶房主要掌管皇帝日常饮茶、烹饮奶茶、运送与保存日常饮用水等。

The room in the northernmost of the east house of Qianqing Palace is the Imperial Tea House, and the plaque was written by emperor Kangxi. It is mainly in charge of the emperors' daily tea drinking, milk tea cooking, delivery and preservation of drinking water.

图2.73　御茶房
Imperial Tea House

清道光 黄地绿龙纹碗

高 6.1、口径 13.7、底径 5.8 厘米

撇口，弧壁，深腹，圈足。内外施黄釉为地，碗心及外壁绘五处绿彩团龙纹，辅"十"字朵云，圈足处绘一周莲瓣纹。圈足内书青花"大清道光年制"六字三行篆书款。

色釉地龙纹瓷器起源于明代永乐时期，清代康熙、雍正、乾隆三朝大量制作。先在白胎上勾出龙纹的图案轮廓，再在白釉上施黄地釉，最后以绿彩绘于刻在胎上的纹饰，经两次低温烧成。据《国朝宫史》记载，黄地绿龙纹器专供清宫贵妃及妃子级别的嫔妃使用。

高 5.9、口径 12.3、底径 4.9 厘米

敞口，深腹，圈足。外壁暗刻双龙戏珠和海水山石纹，内外施黄釉，圈足内书青花"大清光绪年制"六字二行楷书款。黄釉釉色均匀细润，刻绘精细流畅。

颜色釉瓷是历代瓷器尤其是官窑的重要品种，而黄釉瓷在其中占首要地位，黄釉碗是明清时期每朝官窑的必烧品种。

高 4.7、口径 13.4、底径 9.2 厘米

撇口，束颈，垂腹，宽圈足。碗内壁髹金漆，外壁施仿木纹红彩，圈足内施金彩。仿木纹釉瓷是在高温烧成的瓷胎上，以各种色釉涂饰出木材的年轮纹理、树枝疤痕及色彩质感，再经低温烧成，呈现出天然的艺术效果。

清乾隆 碧玉奶茶碗

高 8.2、口径 20、底径 12.6 厘米

由碧玉制成，色碧绿而质厚润。碗口微外撇，深腹，圈足。圈足内刻有"乾隆年制"四字双行隶书款。

清 刻花锡茶叶罐（内装茶叶）

通高 17.5、口径 4.2、底长 13.5 厘米

直口带盖，平肩，扁圆形筒身，平底。盖外錾刻联珠水波纹，罐身錾刻湖石树木纹。罐内装满茶叶。

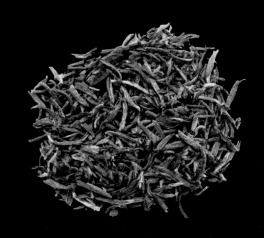

第七章 茶业复兴
——近现代茶文化
Revival of Tea Industry
Modern and Contemporary Tea Culture

从19世纪晚期开始，我国茶叶生产每况愈下。如何振兴华茶成为一个时代课题，让关心茶业命运的近代中国茶人夙夕萦怀。以吴觉农为代表的茶界有识之士进行了一系列改良、改革，使中国的茶业虽然屡经风雨，终不至于凋零。他们为中国茶业保存的薪火和规划的蓝图，至今仍惠及世人。

China's tea production got worse from the late 19th century on, so how to revitalize it became a main concern of the modern Chinese tea experts who cared about the fate of tea industry. Represented by Wu Juenong, the insightful men in the tea circle carried out a series of reforms and innovations, which kept the Chinese tea industry from fading despite repeated trials and tribulations. What they had done for the Chinese tea industry still benefits us today.

一
茶人群像
A Group of Modern Tea Experts

在漫长的中国茶史上，因茶留名者代代皆有。近代以来，又有一群人因为毕生为华茶之复兴而奋斗，而在茶史上留下了不可磨灭的印记。让我们将目光投向这些茶界先贤。

Throughout the long history of Chinese tea, many people have been remembered because of their connection with tea. In modern times, a group of experts strived for the revival of the Chinese tea industry, also leaving an indelible mark in the history of tea. Now let's focus our attention on these great men of the tea circle.

● 吴觉农（1897—1989 年）

浙江上虞人。我国近代茶叶事业的奠基人，在茶叶的生产、贸易、科研、教育等方面提倡和实行一系列科学措施，为中国茶业的振兴奋斗70余年，被人们誉为"当代茶圣"。

● 胡浩川（1896—1972 年）

安徽六安人。茶叶专家，曾任祁门茶叶改良场场长兼财政部贸易委员会皖赣办事处专员、中国茶叶公司总技师等职。中华人民共和国成立后，参与筹建中国茶业公司，主持制订全国茶叶产销计划、茶叶收购加工和出口标准以及加工技术规程。

● 方翰周（1902—1966 年）

安徽歙县人。1935年到江西"宁红"茶区的修水县和"婺绿"茶区的婺源县创建茶叶改良场，并筹建一批实验茶厂和茶场。后参与筹建中南区茶叶公司。中华人民共和国成立后历任中国茶叶总公司产制处处长等职，主持制定茶叶标准、价格、品质。

● 王泽农（1907—1999 年）

江西婺源人。参加筹建复旦大学农学院，任茶叶组教授、茶叶专修科主任；参加筹建财政部贸易委员会茶叶研究所，并任研究员。

● 庄晚芳（1908—1996 年）

福建惠安人。先后任中央大学助教、福建省茶叶管理局副局长、福建省建设局技正、中国茶叶公司研究课课长、福建协和大学教授、复旦大学农学院教授、浙江大学教授、商业部茶叶加工研究所名誉所长等职务。

● 陈椽（1908—1999 年）

福建惠安人。中国制茶学学科的奠基人，现代高等茶学教育事业的创始人之一，为国家培养了大批茶学科技人才，在开发我国名茶生产方面获得了显著成就。著有《制茶全书》《茶业通史》等。

● 冯绍裘（1900—1987 年）

湖南衡阳人。1933年在修水实验茶场负责宁红茶的初、精制试验工作，后赴祁门试制红茶，并在该场设计出一套红茶初制机械设备，开创了我国机制红茶的先例。1938年在云南凤庆创制滇红。被誉为"红茶专家""滇红之父"。

图 2.74　茶人群像
Ten tea experts

二

传薪播火
Education of Tea Science

清末，茶学教育以讲习所或传习所等形式蹒跚起步，虽然课程设置不成体系，招生人数少，教育层次低，却奏响了近现代茶学教育的序曲。

Tea science education started in the form of institute for instruction or training at the end of the Qing Dynasty. Despite the unsystematic curriculum, fewness of enrolled students and low level of schooling of the day, it played the prelude of modern education of tea science.

图2.75 1941年初吴觉农在重庆
Wu Juenong in Chongqing in early 1941

进入中华民国后，各类茶务讲习所和茶业训练班继续发挥作用，造就了一批批在茶叶生产、教育等领域厚植深耕、多有贡献的人才。茶学高等教育也发展起来，其中代表了民国时期茶学教育最高水平的当推复旦大学设立的茶叶系科。

为了满足对茶叶生产和贸易专门人才的需求，在吴觉农的提议和复旦大学的支持下，1940年，内迁重庆的复旦大学成立了中国高等院校中的首个茶叶专业系科，吴觉农出任系主任，授课教师有王泽农、范和钧等人。1940—1946年，复旦大学茶学系共培养了近200名毕业生，为抗战后和中华人民共和国的茶业发展奠定了良好的基础。从此中国的茶学高等教育不断发展，安徽农大、浙江农大、四川农大、湖南农大、华南农大等院校均设立了茶叶系。

In the period of the Republic of China, various tea training institutes and classes continued to play their roles, cultivating professionals who later made great contributions to tea production, education and other fields. Higher education on tea science also developed, and the tea department established by Fudan University represented the highest level of tea science education during this period.

To satisfy the demand for professionals in the field of tea production and trade, in 1940, the first tea science department was established in Fudan University, which was moved inland to Chongqing at that time. Advocating of Wu Juenong and the support of Fudan University contributed to the establishment of the department. Wu Juenong served as the department director. Wang Zenong and Fan Hejun were two of the faculty members. Between 1940 and 1946, nearly 200 persons graduated from the tea department. They were a good foundation for the development of tea industry after the Anti-Japanese War and in People's Republic of China. Since then, China's higher education on tea science has been developing continuously. After the founding of the People's Republic of China, the tea department was set up in many universities, including Anhui Agricultural University, Zhejiang Agricultural University, Sichuan Agricultural University, Hunan Agricultural University and South China Agricultural University.

中国茶学教育诞生于动荡的岁月，展翅于和平的年代。薪火不熄，中国的茶业，未来可期。

China's tea science education started in turbulence and developed in an age of peace. As long as the torch of China's tea industry won't go out, it certainly will have a promising future.

图 2.76　1942 年 6 月复旦大学第一届茶叶专修科毕业合影
Group photo of the first tea specialty graduates of Fudan University in June 1942

图 2.77　抗战时期复旦大学校址
Site of Fudan University during the Anti-Japanese War

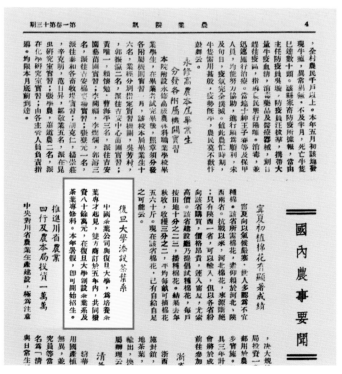

图 2.78　1940 年《农业院讯》刊载复旦大学添设茶叶系的新闻
In 1940, it was published on the *Bulletin of Agricultural Academy* that a tea department would be set up in Fudan University

三

茶叶科技
Tea Science and Technology

为了振兴中国茶业，茶界先贤呼吁在全国主要茶区成立茶业改良场，推广茶叶科技。1932年，吴觉农着手重建祁门茶业改良场，并和胡浩川先后出任场长。祁门茶业改良场、福建的福安茶业改良场和设于三界的浙江省农林改良场茶场等都是当时比较重要的茶业改良实验地。

Experts of the tea circle called for the establishment of improvement farms in the main tea areas in China to promote tea science and technology, in order to revitalize China's tea industry. In 1932, Wu Juenong initiated the rebuilding of the Keemun tea improvement farm, and took the post of director with Hu Haochuan the next one. Farms aimed at tea industry reform in Qimen of Anhui, Fu'an of Fujian and Sanjie of Zhejiang, were important proving grounds for the industry.

民国 "泉祥鸿记茶栈"马口铁茶叶罐

高12.5、长11、宽6厘米

马口铁材质。呈四方体，内嵌式盖。罐身上部彩印"泉祥鸿记茶栈"，主体纹饰为双叶捧门框式布局，内绘远山近水的茶乡风景，意指绿水青山生态佳的茶叶出产地。下部标明茶栈的具体经营地及联系电话。

民国 象牙卷荷龟纹茶则

高 2.5、长 14.5 厘米

象牙材质。此象牙白中泛黄，制作者巧妙地利用象牙原材质设计成卷荷形，背面为阴面，刻画出卷荷的茎叶及蒂柄，阳面设计为卷叶成则。更巧妙的是两只小龟匍匐于卷叶之内，刻画生动，栩栩如生。龟乃灵性之物，我国古人常以龟象征长寿，把茶则设计成卷荷藏龟，大概也同喝茶养生、长寿有关。

茶则是用来量茶入杯的器具，唐代陆羽《茶经》中即已提到茶则。自品茶出现，茶与水的关系历来为人重视，对茶的配量也颇为讲究，量茶入器成为煮茶、点茶及泡茶的重要环节，茶则应运而生。一般茶则以陶瓷、木质、竹质为多，以象牙、犀角、羚角制成的茶则较为少见。

民国 丝茶银行纸币

"拾圆"纸币：纵 9、横 17 厘米
"伍圆"纸币：纵 8.5、横 16 厘米
"壹圆"纸币：纵 8、横 15 厘米

　　三件，纸币面值不用，颜色各异，纹样大体相同。正面中间水印"采茶园"图案，上有"中国丝茶银行"从右向左的排列字样，下有"中华民国十四年"（1925年）的记录，左右两边有红印"天津"字样。背面中心图案为"缫丝图"，上写英文"THE CHINA SILK AND TEA INDUSTRIAL BANK"。中国丝茶银行系由天津巨绅张子青、陈全鼎创办，是一家以发展茶叶、丝绸生产为宗旨的专业性商业银行，总行设在天津，其货币流通于华北地区。

纸质。黑色印刷文字,上写有"股票"字样,内容为"今收到程织文君九如茶号股股本银洋壹百元整,特给股票为证",当时的总理为方树芝,经理为朱永谦,时间为民国十五年(1926年)正月。

中国最早的股票出现于19世纪下半叶,随着近代商业的发展,中国的茶庄、茶号也纷纷采取入股方式筹集资金,此茶号股票正印证了这一史实。

第三章

寻茶问水 咀华啜英
Tea Drinking Practices

一 藏族酥油茶 Tibetan Buttered Tea …… 304
二 维吾尔族奶茶与香茶 Uygur Milk Tea and Fragrant Tea …… 305
三 徽商茶庄 Anhui Merchants' Tea Shop …… 306
四 川蜀茶馆 Sichuan Tea House …… 307

PART III

TEA PROCESSING TECHNIQUES AND DRINKING PRACTICES

技艺

第三篇

第一章

唯有佳茗 不负光阴
A Sip of China and the World … 213

一 余芳千载：历史与现状 Where the Past Meets the Present … 213

二 茶香永续：未来与展望 A Good Smell Well into the Future … 214

第二章

味久而淳 香远益清
The Source of Thrills in Smell and Taste … 215

一 北京 Beijing … 216

二 江苏 Jiangsu … 218

三 浙江 Zhejiang … 224

四 安徽 Anhui … 236

五 福建 Fujian … 244

六 江西 Jiangxi … 258

七 河南 Henan … 264

八 湖北 Hubei … 266

九 湖南 Hunan … 272

十 广东 Guangdong … 278

十一 广西 Guangxi … 280

十二 四川 Sichuan … 284

十三 贵州 Guizhou … 288

十四 云南 Yunnan … 290

十五 陕西 Shaanxi … 303

第一章

唯有佳茗 不负光阴

A Sip of China and the World

> 茶有真香，芳名远播。中华茶文化是中华文明的标识之一，历经数千年的发展，形成了具有地域性、群体性和民族性特征的多样性实践。在传承至今的同时，中华茶文化亦在世界范围内广泛传播，成为世界认识中国的重要途径和中华文明对话世界的桥梁纽带。
>
> Tea is the wit of nature. Tea culture, a significant part of the Chinese civilization, has over millenniums evolved into social practices varying with people of different regions, groups and nations. Now globally widespread, it also facilitates the constant exchanges between China and the rest of the world.

一

余芳千载：历史与现状
Where the Past Meets the Present

1. 茶：中华文明的重要标识

An Iconic Cultural Heritage

茶，是中华文明的象征，是最宝贵的文化财富之一。中国饮茶历史悠久，历经数千年的发展，中国茶形成了绿茶、黄茶、黑茶、白茶、青茶（乌龙茶）、红茶六大基本茶类及品类繁多的再加工茶，两千多种茶品，以不同的色、香、味、形满足着不同群体的需求。中华茶文化具有深厚的文化内涵和地域特色，是中华文明的重要标识和象征。

Tea is a precious heritage. Since long ago that our ancestors started to drink it, now we have had more than 2,000 sorts of tea that, with different colors, smells, tastes and shapes, categorically belong to green tea, yellow tea, dark tea, white tea, oolong tea, black tea and a range of reprocessed teas. Its cultural significance and regional diversity make tea culture a magnificent symbol of Chinese civilization.

2. 茶文化：融于中国人的日常生活
A Life Must for the Chinese

饮茶和品茶贯穿于中国人的日常生活。人们采取泡、煮等方式，在家庭、工作地、茶馆、餐厅、寺院等场所饮用茶与分享茶。在交友、婚礼、拜师、祭祀等活动中，饮茶都是重要的沟通媒介。

Tea is a part of Chinese people's life. Either brewed or boiled, tea has become a popular beverage at home or workplace, or in other places like teahouse, restaurant and temple etc. It's also a good topic at parties, weddings, apprenticeship rites and sacrifices.

二
茶香永续：未来与展望
A Good Smell Well into the Future

1. 非遗传承，使命担当
The Vision of an Intangible Cultural Heritage

家族传承、师徒传承、社区传承和正规教育是中国传统制茶技艺及其相关习俗的主要传承方式。随着多种普及教育措施的实施，青年一代对传统制茶技艺及其保护重要性认识不断提升；相关习俗亦以多样的形式呈现在公众面前，吸引更多人参与其中。

Tea processing techniques and associated practices are usually taught and learned between family members, among a master and his apprentices, by community residents and in schools. The widespread education also helps instill our next generations with a growing sense of tea culture protection and attract more of them to learn and practice associated customs.

2. 源于中国，盛行世界
From China to the World

中国传统制茶技艺及其相关习俗增强了文化认同和社会凝聚力，传达了茶和天下、包容并蓄的理念。以茶敬客、以茶敦亲、以茶睦邻、以茶结友，既增强了社区、群体和个人的认同感和持续感，也是中国人民与世界人民相知相交、中华文明与世界其他文明交流互鉴的重要媒介和人类文明共同的财富。

Traditional tea processing techniques and associated social practices in China have enhanced people's cultural identity and social cohesion, reflecting our quest for a balance between men and nature. As a beverage shared to families, neighbors, friends or even strangers, tea is a bridge to reach the rest of the world and a shared wealth to humankind.

第二章 | 味久而淳　香远益清

The Source of Thrills in Smell and Taste

中国饮茶历史源远流长，制茶技艺传承千年，从古到今众多的茶品犹如灿烂繁星，不可胜数。其制作技艺和技巧以及因此而出现的茶风茶俗更是沉淀为难以统计的文化遗产，凝结成灿若星河的文化景观。截至2021年，共有15个省（自治区、直辖市）的44个茶相关的传统技艺类及民俗类项目入选国家级非物质文化遗产代表性项目名录。这些项目带有鲜明的历史烙印，也是地域文化的缩影，蕴含着弥足珍贵的研究和保护价值。

Tea drinking and its processing skills boasts a history that has spanned through thousands of years in China. The myriads of processing techniques and skills, together with tea customs and practices, have become a spectacular cultural landscape for the country. As of 2021, a total of 44 traditional skills and folk customs from 15 provinces (with autonomous regions or municipalities) have been included into the list of national intangible cultural heritages. The historical and regional characteristics of these elements are priceless in both research and preservation.

北京
Beijing

1. 花茶制作技艺（张一元茉莉花茶制作技艺）
Scented Tea Processing Techniques (Zhang Yiyuan Jasmine Tea)

花茶制作技艺（张一元茉莉花茶制作技艺）主要流布于北京市的张一元茶庄，于2008年列入第二批国家级非物质文化遗产代表性项目名录。

Zhang Yiyuan Jasmine Tea, originated from Zhang Yiyuan Tea House in Beijing, was included into the second batch of national intangible cultural heritage representative list in 2008.

该茉莉花茶制作技艺采用烘青绿茶为茶坯，制作时有10道工序，包括原料验收（茶坯、鲜花检验）、茶坯处理、鲜花处理、茶花拌和、静置窨花、起花、烘焙、茶叶夹杂物处理、提花、匀堆装箱。最终窨制成的茉莉花茶，汤清味浓，入口芳香，回味无穷。

The jasmine tea processing, with baked green tea as the base, comprises 10 steps, namely acceptance (base tea and flower inspection), base tea processing, flower processing, (tea and flower) mixing, scenting, splitting, baking, tea refining, re-scenting and packaging. The jasmine tea smells with a strong flavor and tastes fragrant for long.

图 3.1 茉莉花茶
Jasmine tea

图 3.2 静置窨花
Scenting

图 3.3 采摘茉莉鲜花
Fresh flower picking

图 3.4 鲜花养护
Nurturing

2. 花茶制作技艺（吴裕泰茉莉花茶制作技艺）
Scented Tea Processing Techniques (Wu Yutai Jasmine Tea)

花茶制作技艺（吴裕泰茉莉花茶制作技艺）主要流布于北京市的吴裕泰茶庄，于2011年列入第三批国家级非物质文化遗产代表性项目名录。

Wu Yutai Jasmine Tea, a signature of Wu Yutai Tea House in Beijing, was included in the third batch of national intangible cultural heritage representative list in 2011.

该茉莉花茶制作技艺秉承自采、自窨、自拼原则，工序主要为茶坯制作、花源选择、鲜花养护、玉兰打底、窨制拼和、通花散热、起花、烘焙、匀堆装箱。该技艺在拼配中适当增加茶坯占比，运用"低温慢烘"等方式，形成香气鲜灵持久、滋味醇厚回甘、汤色清澈明亮、叶底经久耐泡的茉莉花茶。

Wu Yutai Jasmine Tea producers pick, scent and blend tea all by themselves. Its processing steps include base-making, flower-selection, flower-nurturing, magnolia-mixing, scenting and blending, ventilating, splitting, baking and packaging. In blending, more base tea is used and techniques like "low-temperature baking" are adopted to instill it with a long-lasting aromatic smell, which also boasts a sweet taste and a clear brew. It is also endurable over several rounds of steeping.

二 江苏 Jiangsu

1. 绿茶制作技艺（碧螺春制作技艺）
Green Tea Processing Techniques (Biluochun)

绿茶制作技艺（碧螺春制作技艺）主要流布于江苏省苏州市的太湖洞庭山一带，于2011年列入第三批国家级非物质文化遗产代表性项目名录。洞庭茶俗称"吓煞人香"，清康熙帝钦定茶名为"碧螺春"。

Biluochun originates from Dongting Mountains off the Taihu Lake in Suzhou, Jiangsu Province, and was included in the third batch of national intangible cultural heritage representative list in 2011. The tea is commonly known as the "startlingly aromatic tea (xia-sha-ren-xiang)". It was renamed "Biluochun" by the Emperor Kangxi of the Qing Dynasty (1644–1911).

图 3.5 碧螺春
Biluochun

图 3.6 冲泡碧螺春
Preparation of Biluochun

碧螺春制作技艺分为采摘、拣剔、摊放、高温杀青、揉捻整形、搓团显毫、文火干燥七道工序，其要领在于"手不离茶，茶不离锅，揉中带炒，炒揉结合，连续操作，起锅即成"。

The production of Biluochun comprises seven steps starting from the picking of tea leaves, which is followed by the selecting, spreading, high temperature fixation, rolling (shaping), rubbing and mild drying. The highlight of this processing technique is that the leaves would be continuously being hand stirred and fried in the cauldron till it gets matured.

其中，"搓团显毫"颇具特色，是形成碧螺春卷曲如螺、茸毫满披的关键。所谓"搓团显毫"，即把茶放在手中搓团，使其出现茸毛。搓团时，锅温控制在50~60℃，时长约12~15分钟，直至茸毫显露，条索细紧。该工序的要点是每搓4~5转后进行一次解块，边搓团，边解块，边干燥。至八成干时，进入下一环节。

The "rubbing" is quite unique and a key to Biluochun's curls like a conch full of white trichomes. "Rubbing" means to rub leaves in hands to make trichomes appear. In rubbing, keep the cauldron temperature at 50–60℃ and rub them for about 12–15 minutes until the trichomes appear visibly straight. The knack rests on unblocking the leaves every 4–5 rubbings. Rub them, unblock them and meanwhile dry them, and go to the next part as tea leaves are 80% dried.

制成的碧螺春条索纤细，卷曲成螺，茸毛遍体，以"形美、色艳、香浓、味醇"四绝闻名。

The processed Biluochun looks slender, curled like spirals and is covered with white trichomes. The tea is known well for its rich color, thick aroma and sweet taste.

① 鲜叶拣剔 Screening fresh teas

② 高温杀青 Fixing at high temperatures

③ 揉捻整形 Rolling and shaping

④ 搓团显毫 Rolling for trichomes

⑤ 文火干燥 Slow-fire drying

图3.7　碧螺春制作工序
The production of Biluochun

2. 绿茶制作技艺（雨花茶制作技艺）
Green Tea Processing Techniques (Yuhua Tea)

绿茶制作技艺（雨花茶制作技艺）主要流布于江苏省南京市，于2021年列入第五批国家级非物质文化遗产代表性项目名录。

Yuhua Tea was invented in Nanjing, Jiangsu Province. It was included in the fifth batch of national intangible cultural heritage representative list in 2021.

雨花茶制作技艺流程分为采摘、摊放、杀青、揉捻、毛火、整形、足火、精制、烘焙、包装。其中整形是集扁形茶类和卷曲形茶类优点于一身的针形茶制作技艺。

The processing of Yuhua Tea comprises picking, spreading, fixing, rolling, initial firing, shaping, full-firing, refining, baking and packaging. Among these steps, shaping is the technique used for the processing of needle shaped tea leaves, having combined the advantages of flat-shaped and curly-shaped tea leaves.

制成的雨花茶形似松针，紧细圆直，锋苗挺秀，色泽绿润，白毫隐露，香气高雅，滋味鲜醇，叶底嫩绿明亮。

Yuhua Tea leaves are needle-like and, as largely in mild green color, are covered faintly with white trichomes. The tea brew sends out a fragrant aroma and tastes mellow. The steeped leaves are usually in a bright green hue.

图 3.8 雨花茶
Yuhua Tea

图 3.9 摊放
Spreading

图3.10 搓条
Twisting

图3.11 抓条理条
Straightening

图3.12 筛分精制
Sieving and refining

3. 茶点制作技艺（富春茶点制作技艺）
Tea Refreshments Processing Techniques (Fuchun Tea Refreshments)

茶点制作技艺（富春茶点制作技艺）主要流布于江苏省扬州市，于2008年列入第二批国家级非物质文化遗产代表性项目名录。

Fuchun Tea Refreshments are the specialty of Yangzhou, Jiangsu Province. It was included in the second batch of national intangible cultural heritage representative list in 2008.

用富春茶点制作技艺制成的点心，造型雅致，品种繁多，甜咸适度，味不雷同。常年制作的品种有三丁包、蟹黄包、青菜包、鲜肉包、豆沙包、干菜包、千层油糕、翡翠烧卖、糯米烧卖等。

Fuchun Tea Refreshments have nice shapes, wide varieties and are known for a balance in taste and flavors. Refreshments that are available all year round include three-ingredient bun, crab-roe bun, veggie-bun, meat bun, bean paste bun, dried-veggie bun, multi-layer oil cake, emerald dumpling, glutinous rice dumpling, etc.

与富春茶点配套的"魁龙珠"茶取安徽魁针之味、浙江龙井之色、江苏珠兰之香，以扬子江水沏泡，有"一壶水煮三省茶"之说。

A tea named "Kuilongzhu", consumed in tandem with the tea refreshments, bears a taste of Kuizhen Tea from Anhui Province, a color of Longjing Tea from Zhejiang Province and a smell of Zhulan Tea from Jiangsu Province. It needs to get steeped with water from the Yangtze River, thus the saying of "a mixed tea from three provinces to get brewed by the same kettle".

图 3.13 "魁龙珠" 茶
Kuilongzhu

图 3.14 茶点制作
Making of tea refreshments

三

浙江
Zhejiang

1. 绿茶制作技艺（西湖龙井）
Green Tea Processing Techniques (West Lake Longjing)

绿茶制作技艺（西湖龙井）主要流布于浙江省杭州市，于2008年列入第二批国家级非物质文化遗产代表性项目名录。

West Lake Longjing comes from Hangzhou, Zhejiang Province. It was included in the second batch of national intangible cultural heritage representative list in 2008.

在长期的生产实践中，西湖龙井产地逐渐摸索出了一套具有鲜明特色的制作技艺，即"抓、抖、搭（透）、拓（抹）、捺、推、扣、甩、磨、压"十大手法。炒制时，先以抓、抖为主；再用搭、抖、捺手法进行初步造型；压的力度由轻而重，压扁成形；起锅摊凉回潮片刻，进行辉锅，开始以理条为主，再转入抓、搭、拓、捺、推、磨等手法并适当加大力度。有经验的制茶师会根据鲜叶嫩度、锅温情况以及锅中茶叶的干燥程度，灵活运用这十种手法。

The tea producers, in centuries of experimentation, acquired a set of distinctive processing techniques, including grasping, shaking, tapping, extending, pressing, pushing, inverting, throwing, grinding and compressing. Take grasping and shaking in frying; and use tapping, shaking and pressing for initial shaping. The compressing strength should be getting heavier and compress leaves into a proper shape. Next, remove tea from the cauldron for cooling and have them moisturized for a short while, and place them into the cauldron again. This time, first straighten leaves and then, apply grasping, tapping, extending, pressing, pushing and grinding. A veteran tea maker would use the skills flexibly according to leaves' tenderness, the cauldron's temperature and the degree of dryness of leaves in the cauldron.

制成的西湖龙井扁平光滑，色泽嫩绿，汤色明亮，嫩栗香或豆花香明显，滋味鲜醇爽口，以"色绿、香郁、味甘、形美"四绝著称，有"绿茶皇后"之美誉。

West Lake Longjing is characterized by its flat shape, tender-green color, bright tea brew, full-bodied chestnuts-like or soybean-milk-like aroma and mellow taste. Its outstanding color, aroma and shape help earn it the title of "the Queen of Green Tea".

图 3.15　西湖龙井茶汤
West Lake Longjing

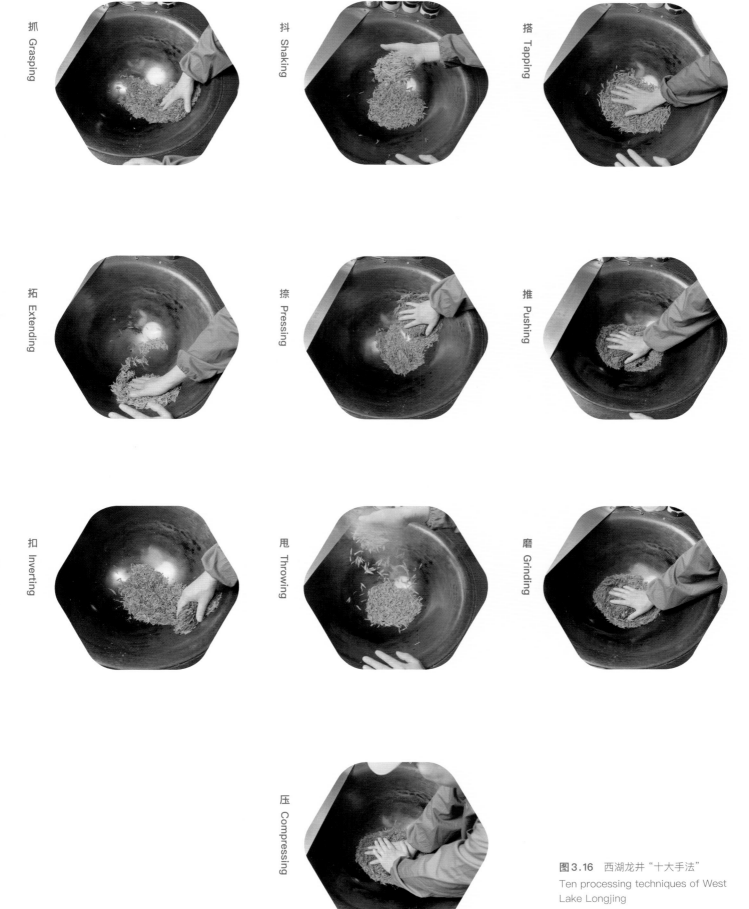

图 3.16 西湖龙井"十大手法"
Ten processing techniques of West Lake Longjing

2. 绿茶制作技艺（婺州举岩）
Green Tea Processing Techniques (Wuzhou Juyan)

绿茶制作技艺（婺州举岩）主要流布于浙江省金华市北山一带，于2008年列入第二批国家级非物质文化遗产代表性项目名录。产地奇峰突兀，怪石嶙峋，有一巨岩耸立，犹如仙人所举，因此而得名。

The processing of Wuzhou Juyan originates from Beishan Mountains in Jinhua, Zhejiang Province. It was included in the second batch of national intangible cultural heritage representative list in 2008. Its place of origin has stacks of grotesque rocks or peaks, among of which there is this one huge rock that looks as if being lifted by an immortal, hence the name Juyan (the rock being lifted).

炒制1千克婺州举岩需要约3万~4万个芽叶。其制作技艺有摊青、青锅、揉捻、二锅、做坯整形、烘焙、精选储存等工序。炒制特点是以炒为主，炒焙结合。

One kilogram of Wuzhou Juyan usually consumes 30,000 to 40,000 buds. The processing techniques comprise spreading, fixing, rolling, re-fixing, shaping, baking and storage. It is largely a fried tea.

图 3.17　摊放
Spreading

婺州举岩茶品细紧略扁，微带茸毫，色泽银翠交辉，香味持久，汤色嫩绿明亮，叶底嫩绿匀整，入口则鲜醇甘美，清新宜人。

The tea is known for its tight, slightly flat shape with visible trichomes and a mixed luster of silvery and green. The prepared tea brew looks bright and green, and it gives out a long-lasting aroma and sweet, refreshing taste.

图 3.18　二锅
Re-fixing

3. 绿茶制作技艺（紫笋茶制作技艺）
Green Tea Processing Techniques (Zisun Tea)

绿茶制作技艺（紫笋茶制作技艺）主要流布于浙江省长兴县，于2011年列入第三批国家级非物质文化遗产代表性项目名录。"紫笋茶"名字由唐代茶圣陆羽《茶经》中"紫者上，绿者次；笋者上，牙者次"的论述得来。

The processing of Zisun Tea prevails in Changxing, Zhejiang Province. It was included in the third batch of national intangible cultural heritage representative list in 2011. The name "Zisun" is derived from a saying of *The Classic of Tea* that "The best tea is purple-colored, followed by the green one, while what is shaped like bamboo shoot (*sun* in Chinese) excels over bud-shaped one".

《茶经》记载："晴，采之、蒸之、捣之、焙之、穿之、封之，茶之干矣。"紫笋茶古法蒸青和制作茶饼的过程，与记载一致。如今，紫笋茶的制作工艺需经过摊青、杀青、回凉、初烘、复烘等14道工序，成品茶芽色带紫，芽形如笋。茶叶舒展后，呈兰花状，尝之齿颊甘香。

According to *The Classic of Tea*, "The tea leaves gathered on a clear day will be steamed, crushed, roasted dry, skewered, and sealed before the end of the day". The procedures of processing Zisun in the past, like steam-fixing and making tea cakes, agreed with the book. Modern processing of Zisun comprises 14 steps, like spreading, fixing, cooling, initial drying and re-baking. With a shade of purple, Zisun Tea looks slender, bamboo-shoot-like in shape. The stretched leaves look like orchids, and the tea brew, long-lasting in fragrance, tastes sweet and mellow.

❶ ❷

图3.19 制作茶饼
Tea cake making

图 3.20 紫笋茶
Zisun Tea

❸

❹

4. 绿茶制作技艺（安吉白茶制作技艺）
Green Tea Processing Techniques (Anji Baicha)

绿茶制作技艺（安吉白茶制作技艺）主要流布于浙江省安吉县，于2011年列入第三批国家级非物质文化遗产代表性项目名录。

Anji, Zhejiang Province, is the home to Anji Baicha's processing. It was included in the third batch of national intangible cultural heritage representative list in 2011.

制作安吉白茶，其工艺可分为采摘、摊放、杀青理条、初烘、摊凉、复烘、收灰干燥七道工序。安吉白茶叶片极薄，对温度、湿度、手法等要求极高，否则易产生焦叶、红梗。

The production of Anji Baicha comprises seven parts, including picking, spreading, fixing and straightening, primary baking, cooling, re-baking and storing for drying. Since its leaves are extremely thin, the process has strict requirements on temperature, humidity and technique, otherwise leaves and stems would be overly burned.

制成的安吉白茶形似凤羽，以白、活、香、鲜、清为特色，鲜叶、干茶均叶白脉绿，颜色鲜活。其氨基酸含量比普通绿茶高3~4倍，茶叶汤色清澈透亮，鲜味足，香气持久。

Shaped like a bird's beautiful, slender feather, Anji Baicha is visibly known for being white, robust, fragrant, fresh and clear. The leaves, either fresh or dried, are white with green veins. The tea's amino acid content 3 to 4 times higher than that of ordinary green tea. The tea soup is clear, with a fresh flavor and long-lasting aroma.

图 3.21 茶园风光
Tea gardens

图 3.22 安吉白茶
Anji Baicha

图 3.23 杀青
Fixing

图 3.24 揉捻
Rolling

图3.25 迎龙灯
Greeting the dragon lamps

5. 庙会（赶茶场）
Temple Fair (Gan-cha-chang)

庙会（赶茶场）主要流布于浙江省磐安县玉山一带，于2008年列入第二批国家级非物质文化遗产代表性项目名录。赶茶场是为纪念晋代"茶神"许逊而举办的活动，后形成以茶叶交易为中心的庙会——"春社"和"秋社"。

Temple Fair (Gan-cha-chang) is a customary practice mostly in Yushan, Pan'an County of Zhejiang Province, and was included in the second batch of national intangible cultural heritage representative list in 2008. Gan-cha-chang is an activity held in memory of Xu Xun, "god of tea" in the Jin Dynasty (265–420). It became later a tea-trading fair specifically staged in spring and fall.

赶茶场包含茶神祭典仪式、民间艺术表演及传统社交活动等。其中，茶神祭典仪式意在保佑茶叶年年丰收、人们平安吉祥。"春社"和"秋社"分别在农历正月十五和十月十五举行，"春社"祭茶神，"秋社"谢茶神。

The Fair is mostly comprised of the Tea God remembrance ritual, folk shows and traditional socializing. The ritual is to pray for tea harvest and people's health. The "Spring Fair" and the "Autumn Fair" are held respectively on the 15th day of the first and tenth months. The "Spring Fair" is to offer sacrifices to the Tea God, while the "Autumn Fair" is to thank him.

图 3.26 迎大旗
The big banner show

图 3.27 祭茶神
Offer sacrifices to the Tea God

图 3.28 舞龙
Dragon dance

6. 径山茶宴
Jingshan Tea Feast

径山茶宴主要流布于浙江省杭州市余杭区，是一种大堂茶会，堪称中国禅茶文化的经典样式。径山茶宴是独特的以茶敬客的庄重礼仪习俗，发源于唐，兴盛于宋，融入僧堂生活和禅院清规，并流传至日本，成为日本茶道之源。

Jingshan Tea Feast is a tea party prevailing around Yuhang District, Hangzhou of Zhejiang Province and is considered a classic style of Chinese Zen tea culture. Jingshan Tea Feast is a unique and solemn ritual in honor of the visiting guests, which originated from the Tang Dynasty and met its prime in the Song Dynasty. The procedures became a part of everyday life of the temple and Zen disciplinary rules. Later it was brought to Japan and became the source of the Japanese Sado.

按照寺里传统，每当贵客光临，住持就在明月堂举办茶宴招待客人。径山茶宴有十多道仪式程序，期间主躬客庄，礼仪备至，依时如法，和洽圆融，体现

图3.29　击茶鼓
Beating drum

图3.30　恭请入堂
Inviting guests into the hall

图 3.31 礼茶敬佛
Burning incense

了禅院清规和礼仪、茶艺的完美结合。其仪式程序有张茶榜、击茶鼓、恭请入堂、上香礼佛、煎汤点茶、行盏分茶、说偈吃茶、谢茶退堂等，期间用"看话禅"的形式问答交谈，机锋偈语，慧光灵现。以茶参禅问道，成为径山茶宴的精髓与核心。

The abbot would usually hold a feast in Mingyue Hall to entertain his guests coming from afar. The ceremonial procedures are composed of a dozen steps, during which both the host and guests have to observe the set rules and etiquette. This perfectly reconcile the Zen philosophy with the tea art. It involves elements like "tea poster" (chabang), drum, incense, tea, verses, etc. till it reaches a formal dismissal. The host and his guests, or the master and his disciples, would be in exchange of verses and wits around Zen issues during the Feast. It's an amazing quest for universal Zen truth from a cup of tea.

图 3.32 煎汤点茶
Making and serving tea in small cups

四

安徽
Anhui

1. 绿茶制作技艺（黄山毛峰）
Green Tea Processing Techniques (Huangshan Maofeng)

绿茶制作技艺（黄山毛峰）主要流布于安徽省黄山市，于2008年列入第二批国家级非物质文化遗产代表性项目名录。

Huangshan, Anhui Province, is the home to the processing of Huangshan Maofeng. It was included in the second batch of national intangible cultural heritage representative list in 2008.

制作黄山毛峰，一般采摘清明至谷雨期间的鲜叶，以一芽一叶为标准，当地称之为"麻雀嘴稍开"。其工艺流程分为鲜叶分级、杀青、摊凉、理条、初烘、摊凉、足烘、提香、揉捻、摊凉、足烘、拣剔等。

Maofeng leaves are usually picked between Qingming and Guyu (roughly in April). A fresh bud must be taken with a leaf, which is described as "a slightly-open sparrow's beak". The processing comprises grading, fixing, cooling, straightening, first baking, cooling, full-baking, slow-baking, rolling, cooling, full-baking and screening.

图3.33 黄山毛峰
Huangshan Maofeng

制成的黄山毛峰条索细扁，形似"雀舌"，带有金黄色鱼叶，色泽嫩绿微黄而油润，俗称"象牙色"，香气清鲜高长，滋味鲜浓醇厚。

Huangshan Maofeng looks slender and flat, like a sparrow's tongue. The leaves usually have earliest golden buds called "yu-ye", which are greenish yellow and glossy, commonly known as "the ivory color". It smells fresh and thick, and tastes refreshing and mellow.

图3.34 茶园风光
Tea gardens

2. 绿茶制作技艺（六安瓜片）
Green Tea Processing Techniques (Lu'an Guapian)

绿茶制作技艺（六安瓜片）主要流布于安徽省六安市，于2008年列入第二批国家级非物质文化遗产代表性项目名录。

The city of Lu'an, Anhui Province, is the home to the processing techniques of Lu'an Guapian. It was included in the second batch of national intangible cultural heritage representative list in 2008.

六安瓜片的技艺流程为鲜叶采摘、拣片、炒片（炒生锅、炒熟锅）、烘焙（毛火、小火、老火）。其中"拉老火"是最后一次烘焙，对形成六安瓜片特殊的色、香、味、形影响极大。此道工序采用木炭烘干茶叶，两人抬烘笼，每笼摊叶约5千克，在炭火上稍作停留后抬走烘笼，轻轻翻动茶叶，然后再次抬上烘笼，历经60~70次。烘至茶叶表面上霜，手捏成粉末，即可下烘。整个过程"火光冲天、热浪滚滚、抬上抬下、以火攻茶"。

The production of Lu'an Guapian comprises picking, sieving, frying (frying raw leaves and frying fried leaves in turn) and baking (on soft, mild and great fire in turn). "Pulling the great fire" is the last baking step that determines tea's color, aroma, taste and shape. In this process, tea leaves are dried by charcoal. The baking cage containing around 5 kilograms of spread leaves will be carried over the charcoal fire for a short while, before bring it away for the stirring of the tea leaves. This will be repeated for 60 – 70 times. Stop baking once the leaves are covered with "frost" and become crispy.

六安瓜片由单片叶制成，不带芽和梗，外形直顺完整，形似瓜子，色泽翠绿，附有白霜。

Lu'an Guapian are made from single leaves without any buds or stems. They are straight and complete, much like melon seeds, and are shining green in color and coated with the white "frost".

图 3.35　六安瓜片
Lu'an Guapian

 ❶ 采摘 Picking
 ❷ 拣片 Sieving
 ❸ 炒生锅 Frying
 ❹ 炒熟锅 Re-frying
 ❺ 拉毛火 Pulling soft fire
 ❻ 拉小火 Pulling mild fire
 ❼ 拉老火 Pulling the great fire

图 3.36 六安瓜片制作工序
The production of Lu'an Guapian

3. 绿茶制作技艺（太平猴魁）
Green Tea Processing Techniques (Taiping Houkui)

绿茶制作技艺（太平猴魁）主要流布于安徽省黄山市黄山区（原太平县），于2008年列入第二批国家级非物质文化遗产代表性项目名录。

The processing of Taiping Houkui, prevalently seen in Huangshan District (formerly known as Taiping County), Huangshan City of Anhui Province, was included in the second batch of national intangible cultural heritage representative list in 2008.

太平猴魁的制作技艺包括采摘、拣尖、摊放、杀青、理条、压制成形、烘茶（头烘、二烘、拖老烘）、装桶。其中，理条是制作太平猴魁特有的工序，该工艺可让芽叶在杀青后保持毫尖完整、叶面舒展、自然挺直。理条时，工匠将出锅的杀青叶一根一根用手捋直、均匀、整齐地铺放于特制的铁纱网盒上。捋直时，要求用手指轻抚，将两片嫩叶包住茶芽，形成"两叶抱一芽"的特征。

The making of Taiping Houkui comprises picking, selecting, spreading, fixing, tidying, compressing and shaping, baking (first, second and third in-depth baking), and packaging. Tidying is a unique step to processing Taiping Houkui. This allows tea buds and leaves to remain intact and the leaf surface stretched and naturally straight after fixing. During the process, tea producers straighten the fixed leaves one by one with hands and lay them evenly and neatly on the special iron mesh box. In straightening, use fingers to gently caress the two young leaves and let them envelope the bud. That's the characteristics of Taiping Houkui.

太平猴魁外形扁展挺拔，色泽苍绿匀润，遍身白毫，主脉呈暗红色，冲泡时芽叶缓慢舒展，竖立成朵，宛如兰花，汤色嫩绿鲜亮，香气鲜灵高爽，滋味鲜爽醇厚。

Taiping Houkui is distinguished for its robust, flat form and delightful green color. The leaves are covered with white hairs and dark red veins. Steeped in water, they will slowly stretch and stand up like an orchid. The brew, tender green and bright, bears an energizing aroma and mellow taste.

图 3.37 太平猴魁
Taiping Houkui

① 采摘 Picking

② 摊放 Spreading

③ 杀青 Fixing

④ 理条 Tidying

⑤ 压制成形 Compressing and shaping

⑥ 烘干（头烘、二烘、拖老烘）Baking (first, second and third in-depth baking)

图 3.38　太平猴魁制作工序
The production of Taiping Houkui

4. 红茶制作技艺（祁门红茶制作技艺）
Black Tea Processing Techniques (Keemun Black Tea)

红茶制作技艺（祁门红茶制作技艺）主要流布于安徽省祁门县，于2008年列入第二批国家级非物质文化遗产代表性项目名录。祁门红茶简称"祁红"，是世界三大高香红茶之一。

The processing of Keemun Black Tea originates from Qimen County, Anhui Province. It was included in the second batch of national intangible cultural heritage representative list in 2008. The black tea, shortly as "Qihong", is one of the world's top three highly-flavored black teas.

祁门红茶制作技艺分为初制和精制两大部分，初制包括萎凋、揉捻、发酵、干燥等工序；精制包括筛分、切断、风选、拣剔、拼配、补火、官堆等工序。筛分有毛筛、抖筛等数道小工序，精工细作。拣剔是剔除各号茶中的破叶、黄片、茶梗和杂物等。补火是将筛拣好的茶叶置于烘笼上，以茶叶烘至褐灰色为适度。官堆是将补火后的各号茶混合倒入官堆中，做成方形高堆，用木齿耙向外梳耙，最终使各号茶混合均匀，即可装箱。

❶ 采摘 Picking　❷ 萎凋 Withering

❺ 干燥 Drying　❻ 筛分 Sieving

图 3.39　祁门红茶
Keemun Black Tea

The processing of Keemun Black Tea comprises primary and refining parts. The former consists of withering, rolling, fermentation and drying, while the latter is comprised of sieving, cutting, winnowing, screening, blending, re-firing and stacking. Sieving has several crafted sub-processes, such as rough sieving and shaking sieving. Screening is to remove broken or wilted leaves, stems and debris. Re-firing is to put the sieved and screened tea leaves on a cage until they turn brownish gray. In the stacking process, each type of re-fired tea leaves will be poured onto the stacking yard and made into a square high stack. Use wooden rakes to rake them outwards to ensure that all types of tea leaves are well mixed, and then package them.

  ❸ 揉捻 Rolling  ❹ 发酵 Fermenting

  ❼ 拣剔 Screening  ❽ 补火 Re-firing

制成的祁门红茶，色泽乌润，条索紧细，香气馥郁持久，被誉为"祁门香"。

Keemun Black Tea looks dark-red in color, slender in form and a lingering and high aroma, thus being praised as "Keemun Fragrance".

 ❾ 官堆 Stacking

图3.40 祁门红茶制作工序
The production of Keemun Black Tea

五 福建 Fujian

1. 白茶制作技艺（福鼎白茶制作技艺）
White Tea Processing Techniques (Fuding Baicha)

白茶制作技艺（福鼎白茶制作技艺）主要流布于福建省福鼎市，于2011年列入第三批国家级非物质文化遗产代表性项目名录。茶品因芽头肥壮，满披白毫，如银似雪而得名。

The processing of Fuding Baicha, now mostly found in Fuding, Fujian Province, was included in the third batch of national intangible cultural heritage representative list in 2011. The tea is so named for its plump and hair-covered buds so white as silver and snow.

福鼎白茶制作技艺不炒不揉，文火足干，以适度的自然氧化保留丰富的活性酶和多酚类物质。其初制工艺流程主要包括萎凋、堆积、干燥和拣剔。精制工艺流程为拣剔（手拣）、正茶、匀堆、烘焙和装箱。

No frying or rolling, Fuding Baicha simply requires slow drying. Such moderate natural oxidation can help retain a rich supply of active enzymes and polyphenols. The primary processing comprises withering, stacking, drying and screening, while the refining processing is comprised of screening (hand-sort), framing, stacking, baking and packaging.

萎凋是福鼎白茶制作技艺的关键工序。明代《煮泉小品》载："茶者，生晒者为上，亦更近自然。"其中的日晒干燥法被认为是白茶制作技艺的雏形。当代的萎凋技艺主要分为自然萎凋和复式萎凋两种方式。自然萎凋，即将鲜叶放置在通风的晾青架上进行萎凋。复式萎凋则是将茶叶放在日光下进行自然萎凋并结合室内萎凋，且每次日照时间不超过半小时。

Withering is the key part for processing Fuding Baicha. *Essays on Tea & Water* back to the Ming Dynasty noted that "The fire-dried tea is second to the sun-dried tea, as the latter seems to be more natural". The said sun-drying is considered the prototype of white tea processing techniques. Today, withering is mainly divided into natural withering and compound withering. The former has fresh tea leaves placed on a ventilated drying rack for withering, and the latter requires them to get naturally withered in the sunlight and in rooms as well (the sunlight time shall not exceed half an hour each time).

根据不同的树种和采摘标准，可以制成白毫银针、白牡丹、贡眉（寿眉）以及新工艺白茶等不同品种的福鼎白茶。

Based on tea plant species and picking standards, Baihao Yinzhen (Silver Needles), Baimudan (White Peony), Gongmei (or Shoumei) and "new bai-cha" are produced as the family members of Fuding Baicha.

图 3.41 福鼎白茶
Fuding Baicha

❶ 采摘 Picking

❷ 萎凋 Withering

❸ 堆积 Stacking

❹ 干燥 Drying

❺ 拣剔 Screening

图 3.42　福鼎白茶制作工序
The production of Fuding Baicha

图 3.43 采茶
Picking tea leaves

2. 武夷岩茶（大红袍）制作技艺
Wuyi Rock Tea (Dahongpao) Processing Techniques

武夷岩茶（大红袍）制作技艺主要流布于福建省武夷山一带，于2006年列入第一批国家级非物质文化遗产代表性项目名录。

The processing of Wuyi Rock Tea (Dahongpao), mostly seen in the area of Mount Wuyi, Fujian Province, was included in the first batch of national intangible cultural heritage representative list in 2006.

武夷岩茶（大红袍）制作技艺由绿茶、红茶制作技艺演变而成，基本制作流程包括采摘、萎凋、做青、炒青与揉捻（双炒双揉）、初焙（俗称"走火焙"）、扬簸、晾索、拣剔、复焙、团包、补火、毛茶装箱。

The processing techniques actually evolved from those of green tea and black tea. The process basically covers picking, withering, rotating, frying and rolling (twice), primary baking (commonly known as "on-fire baking"), tossing, drying, screening, re-baking, wrapping, re-firing and packaging.

图3.44　武夷岩茶（大红袍）
Wuyi Rock Tea (Dahongpao)

做青是武夷岩茶（大红袍）制作技艺中形成"绿叶红镶边"的过程。该工艺包括摇青和做手。摇青时，青叶在水筛内呈螺旋形，上下翻滚，叶缘之间互相碰撞摩擦。在此期间，用手掌轻拍茶青，弥补摇青力度的不足（俗称"做手"）。做青过程中，可将青叶静置几次。做青须"看天做青，看青做青"，持续时间约8～10小时，最后一次做青时，可看到叶面凸起呈龟背形（俗称"汤匙叶"），红边显现。

Rotating is a process to generate "red edges against the green surface" for Dahongpao. It consists of hand shaking and patting. In hand shaking part, tea leaves will be rolled up and down in water sieve, causing the edges to collide and rub against each other. They will also be gently hand-patted to enhance the effect. When shaking is done, tea leaves are to be left alone for a short period of time. Weather conditions also matter for the process. Each shaking part will take 8 – 10 hours. During the final shaking, patterns of turtle's back (commonly known as "spoon leaf") and the red edges will simultaneously start to appear on the surface of tea leaves.

制成的武夷岩茶（大红袍）具有"岩骨花香"的品质，其外形呈长条眉状，汤色呈琥珀色，叶底明亮，香气清幽浓长，滋味醇厚鲜爽。

Dahongpao has "a rock-and-flower-like fragrance". It looks slender and eyebrow-like, and the brew with clean post-brewing leaves inside looks amber-like, and has a long standing fragrance and thick, mellow taste.

❶ 萎凋 Withering

❹ 初焙 Primary baking

❼ 拣剔 Screening

❷ 摇青 Shaking

❸ 炒青与揉捻 Frying and rolling

❺ 扬簸 Tossing

❻ 晾索 Drying

❽ 复焙 Re-baking

❾ 装箱 Packaging

图 3.45　武夷岩茶（大红袍）制作工序
The production of Wuyi Rock Tea(Dahongpao)

3. 乌龙茶制作技艺（铁观音制作技艺）
Oolong Tea Processing Techniques (Tieguanyin)

乌龙茶制作技艺（铁观音制作技艺）主要流布于福建省安溪县，于2008年列入第二批国家级非物质文化遗产代表性项目名录。

The processing of Tieguanyin, mostly popular across Anxi County, Fujian Province, was included in the second batch of national intangible cultural heritage representative list in 2008.

在铁观音制作技艺中的初制包含晒青、晾青、摇青、炒青、揉捻、初烘、包揉、复烘、复包揉、烘干十道工序。精制包含筛分、拣剔、拼堆、烘焙、摊凉、包装六道工序。

The primary processing of Tieguanyin comprises 10 steps, including sun-drying, air-drying, shaking, frying, rolling, first baking, wrapping and rolling, re-baking, re-wrapping and re-rolling, and roast-drying, while the refining processing comprises sieving, screening, stacking, baking, cooling and packaging.

摇青又称"筛青"，是铁观音制作技艺的关键工序，可形成"绿叶红镶边"现象。摇青时，使用悬挂的半球形大竹筛，即"吊筛"，离地高度以方便使用为准。每次摇青的投叶量为5~6千克。制茶师握住筛沿作前后往复、上下簸动，使叶子在茶筛中呈波浪式翻滚。在"筛动"与"静凉"的反复中，叶子不断出现"退青"和"还阳"现象。摇青历时10~12小时，直至出现青蒂、绿腹、红镶边。

Shaking is also known as "sieving", and it plays a key part in making Tieguanyin to adorn green leaves with red rims. A large suspended hemispherical bamboo sieve, known as "hanging sieve", is used in the process. The height of the sieve from the ground can be adjusted based on requirements. The amount of leaves to be shaken each time is 5–6 kilograms. Shake the sieve back and forth, up and down by holding the sieve edge, to roll and get them flipped over. Through repeated "sieving" and "cooling", the green shade of tea leaves will appear and fade away alternatively. Keep it done for 10–12 hours, and you will find the stems and leaves' back surface green, and red rims on the green leaves.

制成的茶叶乌润结实，沉重似铁，并带有天然的"兰花香"和特殊的"观音韵"。

Tieguanyin has dark, heavy leaves. It has naturally an "orchid-like fragrance" and a special "Guanyin (Kwan-yin) taste".

图 3.46　铁观音
Tieguanyin

❶ 晒青 Sun-drying

❷ 摇青 Shaking

❸ 炒青 Frying

❹ 揉捻 Rolling

❺ 初烘 First baking

❻ 包揉 Wrapping and rolling

图 3.47 铁观音制作工序
The production of Tieguanyin

4. 乌龙茶制作技艺（漳平水仙茶制作技艺）
Oolong Tea Processing Techniques (Zhangping Shuixian)

乌龙茶制作技艺（漳平水仙茶制作技艺）主要流布于福建省龙岩市，于2021年列入第五批国家级非物质文化遗产代表性项目名录。

The processing of oolong tea (Zhangping Shuixian) originates from Longyan, Fujian Province. It was included in the fifth batch of national intangible cultural heritage representative list in 2021.

漳平水仙茶制作技艺是乌龙茶类中的紧压茶技艺，吸取了武夷岩茶和闽南水仙茶的制作原理，技艺流程为晒青、晾青、做青（摇青和静置）、炒青、揉捻、毛拣、模压造型、烘焙、摊凉等。

Zhangping Shuixian is a sort of compressed oolong tea, whose processing has referred to the techniques of processing Wuyi Rock Tea and Shuixian Tea of the southern Fujian. The processing comprises sun-drying, air-drying, shaking (hand shaking and standing), frying, rolling, screening, molding, baking and cooling.

模压造型是漳平水仙茶制作技艺的特有工序。模压造型的工具主要为特制的木模和木模槌。操作时，将毛边纸或热封型滤纸平铺于桌面上，上置内边为4厘米×4厘米的木模，加入约14克揉捻叶，再用木槌加压造型，成形后将纸包扎紧，用米浆粘封。

For it processing, molding is a unique part. The tools for molding are special wooden molds and pestles. In molding process, the deckle-edged paper or heat-sealed filter paper will first be laid flat on the table with a wooden mold (4*4 cm) placed on its top. Put 14 grams of rolled leaves into the mold, get them compressed and shaped with the pestle. Then, wrap the tea with paper and seal it with rice pulp.

制成的茶品形似方饼，具有天然的"兰花香"或"桂花香"。

Like a square tea cake, Zhangping Shuixian bears a fragrance like both orchid and sweet osmanthus.

图 3.48 漳平水仙茶
Zhangping Shuixian

❸ 炒青 Frying   ❹ 揉捻 Rolling

❼ 烘焙 Baking

图3.49　漳平水仙茶制作工序
The production of Zhangping Shuixian

图3.50　模压工具
Molding tools

5. 红茶制作技艺（坦洋工夫茶制作技艺）
Black Tea Processing Techniques (Tanyang Congou Tea)

红茶制作技艺（坦洋工夫茶制作技艺）主要流布于福建省福安市白云山麓一带，于2021年列入第五批国家级非物质文化遗产代表性项目名录。

The processing of Tanyang Congou Tea, mostly found at the foot of Mount Baiyun in Fu'an, Fujian Province, was included in the fifth batch of national intangible cultural heritage representative list in 2021.

制作坦洋工夫茶，首先需经过萎凋、揉捻、发酵、干燥等工序，做成条索状的红毛茶；再经过筛分、切断、风选、拣剔、复火、匀堆等工序，分成各级的精制茶。

The first half of the processing includes withering, rolling, fermenting and drying, aims to make crude black tea strips. The second half comprises sieving, cutting, winnowing, screening, re-baking and stacking, after which the tea will become refined tea of various grades.

制成的茶品外形细长匀整，带白毫，色泽乌黑，内质香味清鲜，汤色明亮呈金黄色，有桂花的浓郁芳香，滋味鲜浓、醇厚。

Tanyang Congou Tea looks slender and neat, appears in dark shade, and is covered with white trichomes. The gold-colored tea brew smells like sweet osmanthus in full blossom and tastes fresh, thick and mellow.

图 3.51　坦洋工夫茶
Tanyang Congou Tea

图3.52 筛分
Sieving

图3.53 匀堆
Stacking

6. 花茶制作技艺（福州茉莉花茶窨制工艺）
Scented Tea Processing Techniques (Scenting Processing Techniques of Fuzhou Jasmine Tea)

花茶制作技艺（福州茉莉花茶窨制工艺）主要流布于福建省福州市，于2014年列入第四批国家级非物质文化遗产代表性项目名录。

The processing of Fuzhou Jasmine Tea are mostly seen in Fuzhou, Fujian Province. It was included in the fourth batch of national intangible cultural heritage representative list in 2014.

福州茉莉花茶窨制工艺流程主要包括茶坯粗制、伺花和精制、茶花拼和（窨花）、静置通花、收堆复窨、茶花分离（起花）、烘焙、转窨或提花、匀堆装箱。窨制是福州茉莉花茶窨制工艺的重点工序，又称"窨花"，即茶花拼和。窨制对配花量、花开放程度、温度、水分、厚度、时间有较高要求，窨次不同，对这六要素的要求也不同。制茶师将一层花、一层茶重重叠叠堆放，充分拌匀和通氧，让花不失生机，茶吸收新鲜的花香达到饱和状态，最终形成茶味与花香融和无间的特点。

The processing of Fuzhou Jasmine Tea covers base-processing, refining, mixing (scenting), ventilating, re-collecting and re-scenting, flower-tea separation, baking, final scenting, even stacking and packaging. Scenting is a key part of processing Fuzhou Jasmine Tea, in fact a blending of tea and flowers. Scenting is particular about the quantity of flowers, degree of flowering, temperature, moisture, thickness and time, and the requirements can vary from time to time. Mix flowers and tea in layers and get them ventilated to keep flowers alive as long as possible, which allows tea to absorb the fresh aroma of flowers to the fullest. The jasmine tea can actually be a blend of floral fragrances.

图 3.54 福州茉莉花茶
Fuzhou Jasmine Tea

图 3.55 茶花拌和
Blending tea and flowers

图 3.56　筛花
Sieving flowers

六

江西
Jiangxi

1. 绿茶制作技艺（赣南客家擂茶制作技艺）
Green Tea Processing Techniques (Hakka Lei Cha in Southern Jiangxi Province)

绿茶制作技艺（赣南客家擂茶制作技艺）主要流布于江西省全南县客家人居住的区域，于2014年列入第四批国家级非物质文化遗产代表性项目名录。

The processing of Hakka Lei Cha, prevalently seen in the Hakka areas of Quannan County, Jiangxi Province, was included in the fourth batch of national intangible cultural heritage representative list in 2014.

赣南客家擂茶制作技艺中使用的基本原料包括鲜茶叶、糯米、芝麻、黄豆、花生、盐及各类青草药等。制作时，擂茶者多为端坐，双腿夹住擂钵，手握擂杵沿钵内壁顺沟纹走向频频舂捣、旋转磨成茶泥；再将擂好的茶泥放在擂钵或大盆里，用滚烫的开水冲泡；最后加入些许高山茶油，用擂杵搅拌，甘润芳香、色如琥珀的擂茶便制作完成了。

The ingredients of Hakka Lei Cha are fresh tea leaves, glutinous rice, sesame seeds, soybeans, peanuts, salt and various medicinal herbs. In processing, sit upright with legs clamping the bowl, hold the pestle and pound the mixture following the direction of patterns on the inner wall of the bowl. Next, put the smashed pastes into a bowl or a larger vessel and steep them with boiling water. Finally, mix them with a little alpine tea oil and stir the paste with the pestle for a while, and here is the amber-like and fragrant Lei Cha.

制成的擂茶清爽可口，既有茶叶的甜味，芝麻、花生、豆子的清香，也有生姜的辣味，口感极为丰富。

Lei Cha emits a sweet and refreshing aroma and tastes a mixed flavor of tea, sesame, peanuts, beans and ginger.

图 3.57　擂茶原料
Ingredients of Lei Cha

图 3.58 制作擂茶
Making Lei Cha

2. 绿茶制作技艺（婺源绿茶制作技艺）
Green Tea Processing Techniques (Wuyuan Lvcha)

绿茶制作技艺（婺源绿茶制作技艺）主要流布于江西省婺源县，于2014年列入第四批国家级非物质文化遗产代表性项目名录。

The processing of Wuyuan Lvcha (green tea), prevalently found in Wuyuan County, Jiangxi Province, was included in the fourth batch of national intangible cultural heritage representative list in 2014.

图 3.59　婺源绿茶
Wuyuan Lvcha

婺源绿茶制作技艺的特征为：小锅杀青，小桶揉捻，干燥低温长焙。技艺流程分为采摘、摊片、杀青、揉捻（解块）、做形（初干）、烘干（做香）。揉捻分为冷揉和热揉，冷揉在杀青叶经摊凉后进行，热揉则是杀青叶不经摊凉而趁热进行。

In Wuyuan Lvcha processing, green tea leaves are first fixed and rolled in small buckets, then dried by a low heat. The processing comprises picking, spreading, fixing, rolling (unblocking), shaping (first drying) and baking (scenting). The rolling can be divided into cool rolling and hot rolling. The former refers to the rolling of tea leaves after they are dried and cooled, while the latter means to roll the tea leaves while they are in the process of drying.

制成的婺源绿茶色泽润绿，外形紧细圆直，汤色清澈，滋味鲜爽，被赞为"头泡香、二泡浓、三泡味未减、四泡味亦醇"。

The green tea is lustrously green in color and has a clear, fragrant and refreshing brew. As the saying goes, "It will smell fragrant on the first brew, full-flavored on the second, and aroma will still be lingering on the third and fourth".

图 3.60　揉捻
Rolling

图 3.61 宁红茶
Ninghong Tea

3. 红茶制作技艺（宁红茶制作技艺）
Black Tea Processing Techniques (Ninghong Tea)

红茶制作技艺（宁红茶制作技艺）主要流布于江西省修水县，于2021年列入第五批国家级非物质文化遗产代表性项目名录。

The processing of Ninghong Tea, prevalently adopted in Xiushui County, Jiangxi Province, was included in the fifth batch of national intangible cultural heritage representative list in 2021.

宁红茶制作技艺包括毛茶、毛火（熟做）、筛制、拣剔、补火、官堆、装箱。其制作优势是生产安全，制成的茶叶能较好发挥原料的经济价值。

The processing of Ninghong Tea comprises crude tea making, pre-firing (cooking), sieving, screening, re-firing, stacking and packaging of the tea leaves. The procedures are safe in tea production and can bring the tea's economic value to the full.

图 3.62 揉捻
Rolling

制成的宁红茶条索紧结匀整，锋苗挺拔，色泽黝黑油润，汤色浓艳透明，滋味鲜浓甜爽。除了散条形茶外，宁红茶还会制成一种捆扎茶——龙须茶，因其叶条似须而得名。

The appearance of Ninghong Tea is neat, slender and well-proportioned and is in lustrous black color. Its bright-colored brew packs a strong, refreshing flavor. Besides the loose leaves form, there is a bundled package of Ninghong called Longxu Tea (tea like dragon's whiskers), which was so named because it bears a certain resemblance to the whiskers of the mystical Chinese dragon.

图 3.63　信阳毛尖茶
Xinyang Maojian

七

河南
Henan

绿茶制作技艺（信阳毛尖茶制作技艺）
Green Tea Processing Techniques (Xinyang Maojian)

绿茶制作技艺（信阳毛尖茶制作技艺）主要流布于河南省信阳市，于2014年列入第四批国家级非物质文化遗产代表性项目名录。

The processing of Xinyang Maojian, widely seen in Xinyang, Henan Province, was included in the fourth batch of national intangible cultural heritage representative list in 2014.

信阳毛尖茶制作技艺主要包括分级、摊放、生锅、熟锅、初烘、摊凉、复烘、毛茶整理、提香。其中，"生锅"杀青和"熟锅"甩条手法是两大特色。

The production technique of Xinyang Maojian mainly includes grading, spreading, primary firing, secondary firing, initial drying, spreading and cooling, re-drying, tea leaf finishing, and scent enhancement. Among them, the "primary firing" for fixation and the "secondary firing" for shaping are two major features.

制成的信阳毛尖茶细圆挺秀，满毫匀齐，色泽翠绿，茶水以汤清色绿、香高味浓耐泡、滋味醇香为佳。

The appearance of Xinyang Maojian is slender and is evenly covered with trichomes amidst the emerald shade. The brew is clear with a high flavored aroma, and the tea tastes thick and mellow.

图 3.64 生锅
Primary firing

图 3.65 烘焙
Baking

八

湖北
Hubei

1. 绿茶制作技艺（恩施玉露制作技艺）
Green Tea Processing Techniques (Enshi Yulu)

绿茶制作技艺（恩施玉露制作技艺）主要流布于湖北省恩施市一带，于2014年列入第四批国家级非物质文化遗产代表性项目名录，是我国少有的蒸青针形绿茶技艺。

The processing of Enshi Yulu, prevalent in Enshi, Hubei Province, was included in the fourth batch of national intangible cultural heritage representative list in 2014. It is a rather rare type of needle-shaped steamed green tea in China.

图 3.66　恩施玉露
Enshi Yulu

❶ 蒸青 Steaming

❷ 扇干水汽 Drying

❺ 铲二毛火 Re-frying

❻ 整形上光 Shaping

恩施玉露制作技艺流程包括鲜叶摊放、蒸青、扇干水汽、炒头毛火、揉捻、铲二毛火、整形上光、焙火提香、拣选，包含蒸、扇、抖、揉、铲、整六大核心技术和搂、端、搓、扎四大手法。

The processing comprises the spreading of fresh leaves, steaming, drying, initial frying, rolling, re-frying, shaping, baking and screening. It features steaming, fanning, shaking, rolling, shovelling and shaping as the key six skills, and raking, tidying, rubbing and binding as four major techniques.

蒸青是恩施玉露制作技艺的特色工艺，保留了唐宋时期的特色。蒸青要求高温、薄摊、短时、快速。将蒸青盒插入蒸青箱内，待盒内温度接近100℃，迅速把鲜叶均匀薄摊于盒内。蒸青时间一般为30秒，较老叶子适当延长时间，以鲜叶失去光泽、叶质柔软、青气消失、茶香显露为适度。

Steaming is a special part in processing Enshi Yulu. This is an age-old technique that has retained characteristics back to the Tang and Song Dynasties. Steaming requires a high temperature, thin spreading, short time and fast operations. First, insert the steaming box into the steaming case, and then evenly and thinly spread tea leaves in the box quickly as long as the temperature inside the box is close to 100℃. Steam it for 30 seconds (or longer for older leaves) till the fresh leaves lose the luster and become soft, together with the missing of its floral smell and the birth of tea fragrance.

制成的茶品色泽润绿，条索匀整光滑，挺直如松针，香高持久；冲泡后汤色嫩绿明亮，滋味鲜醇，叶底绿亮匀整。

Enshi Yulu is lustrous green in color, having a straight, pine-needle-like shape. With a delicate and lasting aroma, the brewed tea looks clear and light and has a mellow flavor.

❸ 炒头毛火 Initial frying

❹ 揉捻 Rolling

❼ 焙火提香 Baking for a stronger fragrance

图3.67　恩施玉露制作工序
The production of Enshi Yulu

2. 黑茶制作技艺（赵李桥砖茶制作技艺）
Dark Tea Processing Techniques (Zhaoliqiao Brick Tea)

黑茶制作技艺（赵李桥砖茶制作技艺）主要流布于湖北省赤壁市，于2014年列入第四批国家级非物质文化遗产代表性项目名录。赵李桥砖茶的前身是羊楼洞砖茶，羊楼洞是茶马古道的源头之一。

The processing of Zhaoliqiao Brick Tea, prevalently seen in Chibi, Hubei Province, was included in the fourth batch of national intangible cultural heritage representative list in 2014. Zhaoliqiao Brick Tea was previously known as Yangloudong, which was a source of "the Old Tea-Horse Road".

赵李桥砖茶制作技艺可以制成青砖茶和米砖茶两类。制作青砖茶，先将鲜叶制成毛茶，其面茶技艺分杀青、初揉、初晒、复炒、复揉、渥堆、晒干等；里茶技艺分杀青、揉捻、渥堆、晒干四道工序，再精制成茶品。米砖茶原料为末茶，工序为筛分、拼料、压制、退砖、检砖、干燥等。

Using the Zhaoliqiao Brick Tea processing techniques, tea can be made into either black or rice brick tea. As fresh leaves are made into coarse tea, the processing "mian-cha" involves fixing, first rolling, first drying, re-frying, re-rolling, stacking and drying, while in making "li-cha", it requires fixing, rolling, stacking and drying. Rice brick tea is made of tea grounds and their processing techniques include sieving, mixing, compressing, brick shaping, brick-checkup and drying.

制成的茶品外形美观，砖模棱角分明，纹面图案清晰秀丽。

The brick tea is sharp-edged, fine-looking with clear, sophisticated designs.

图 3.68 青砖茶
Black Brick Tea

图 3.69 晒茶
Sun-drying

图 3.70 米砖茶
Rice Brick Tea

图 3.71 米砖茶
Rice Brick Tea

3. 黑茶制作技艺（长盛川青砖茶制作技艺）
Dark Tea Processing Techniques (Changshengchuan Brick Tea)

黑茶制作技艺（长盛川青砖茶制作技艺）主要流布于湖北省宜昌市，于2021年列入第五批国家级非物质文化遗产代表性项目名录。

The processing of Changshengchuan Brick Tea, prevalently seen in Yichang, Hubei Province, was included in the fifth batch of national intangible cultural heritage representative list in 2021.

长盛川青砖茶制作技艺从采摘到包装共77道工序，包括杀青、渥堆、干燥、陈化、筛分、精制、蒸压、定形、烘干等。渥堆是黑茶制作技艺中的特有工序，也是形成黑茶品质、口感、香气的关键性工序。渥堆时，将茶叶层层铺在发酵池中，在湿热作用下，与空气中的微生物发生自然酶促反应，使茶叶内生物质进行发酵转化。

图 3.72　长盛川青砖茶
Changshengchuan Brick Tea

From picking to packaging, the entire process of Changshengchuan Brick Tea comprises 77 steps, including fixing, stacking, drying, aging, sieving, refining, steaming, shaping and baking. Stacking is a particularly unique to the

❶ 杀青 Fixing
❷ 渥堆 Stacking
❸ 翻堆 Overturning the stack

❻ 蒸茶 Steaming
❼ 入模 Mold insertion
❽ 压制 Compressing

dark tea, and also a tea process to form the quality, taste and aroma of dark tea. When the tea is stacked, the tea leaves are spread layer by layer in the fermentation pool. Under the action of wet and heat, the natural enzymatic reaction takes place with the microorganisms in the air to ferment and transform the biomass in the tea.

在长盛川青砖茶制作技艺中，茶料经精制、拼配后，便可进入定形工序。紧压设备为木制牛皮架，是使用了二级杠杆原理的传统工具。操作时，须由7~8名制茶师分工协作，在牛皮架两侧分别施行备料、称茶、蒸茶、装斗、置面板、加衬板、降扳机、压茶、关卡、升扳机、出斗、退砖、修砖等工序。紧压后的茶砖，内部茶叶与空气隔绝，茶叶氧化趋于停滞，再经过烘干、包装后即可。

Tea will be shaped after being refined and blended. People use a wooden cowhide frame, a traditional tool applying the secondary lever principle. It requires 7 – 8 tea makers to work together, feeding the machines, weighing and steaming tea leaves, and loading the bucket before they are compressed into bricks. The tea bricks, as being airtight inside, need to be dried and packaged.

制成的茶品经历了散茶、柱子茶、饼茶、砖茶等多种形态，色泽青褐，纹饰清晰，香气纯正，汤色红黄尚明。

The made tea will appear in various forms, such as loose tea, pillar tea, cake tea, brick tea, etc. The tea looks black and brown in color, and with clear patterns and a strong fragrance. The brew is orange.

摊晾干燥 Spreading and drying

风选 Winnowing

烘干 Baking

图 3.73　长盛川青砖茶制作工序
The production of Changshengchuan Brick Tea

九

湖南
Hunan

1. 黄茶制作技艺（君山银针茶制作技艺）
Yellow Tea Processing Techniques (Junshan Yinzhen)

黄茶制作技艺（君山银针茶制作技艺）主要流布于湖南省岳阳市君山区，于2021年列入第五批国家级非物质文化遗产代表性项目名录。

The processing of Junshan Yinzhen, prevalent in Junshan District, Yueyang City of Hunan Province, was included in the fifth batch of national intangible cultural heritage representative list in 2021.

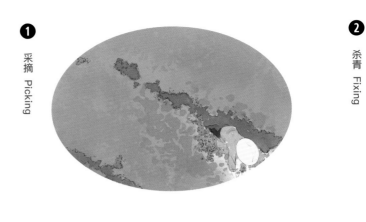

君山银针茶制作技艺分采摘、摊晾、杀青、摊凉、初烘、初包、复烘、复包、足火、精选十道工序，制茶历时约4天。闷黄是君山银针茶制作技艺的重要工序，在此过程中，黄茶会形成特有的色、香、味。初包闷黄，是将初烘叶用牛皮纸包好，每包约1.5千克，在箱内放置40~48小时。期间温度上升，应进行翻包，以使转色均匀。复包主要是补充初包发酵程度的不足，以利于形成黄茶特有的品质。复包历时约20小时，至茶芽色泽金黄、香气浓郁为适度。

图 3.74 君山银针茶
Junshan Yinzhen

The production of Junshan Yinzhen, which takes up to 4 days in total, comprises ten steps, namely picking, spreading, fixing, cooling, first baking, first wrapping, re-baking, re-wrapping, full-firing and screening. Yellowing is an important part in processing Junshan Yinzhen, in which yellow tea will develop its unique color, aroma and taste. The first wrapping means to wrap the baked leaves with kraft paper (1.5 kg each pack) and place them in an odorless wooden box for 40–48 hours for "yellowing". Over this period, as the temperature rises, tea packs are overturned to ensure even yellowing. Re-wrapping is to enhance the less-fermented leaves from the first wrapping to generate the unique quality of yellow tea. Re-wrapping lasts for about 20 hours, until the tea buds are golden in color and rich in aroma.

❸ 摊凉 Cooling

❹ 初烘 First baking

❼ 复包闷黄 Re-wrapping for yellowing

❽ 足火 Full-firing

制成的君山银针茶外形芽壮多毫，带有淡黄色茸毫，有"金镶玉"之誉。冲泡后，香气清鲜，汤色浅黄，滋味甜爽，叶底明亮。

Junshan Yinzhen features robust sprouts and buds that are evenly coated with pale-yellow trichomes, known as "gold inlaid with jade". The prepared light-yellow brew gives off a refreshing aroma with post-brewing leaves being clear, and it has a sweet and brisk taste.

❾ 精选 Screening

图3.75　君山银针茶制作工序
The production of Junshan Yinzhen

2. 黑茶制作技艺（千两茶制作技艺）
Dark Tea Processing Techniques (Qianliang Cha)

黑茶制作技艺（千两茶制作技艺）主要流布于湖南省安化县，于2008年列入第二批国家级非物质文化遗产代表性项目名录。清代茶商将千两茶捆踩成圆柱状，每只定为老秤的一千两，故得名。

The processing of Qianliang Cha, widely adopted in Anhua County, Hunan Province, was included in the second batch of national intangible cultural heritage representative list in 2008. During the Qing Dynasty, the tea was bundled into a cylindrical shape with each bundle being 1,000 *liang* (a Chinese weight unit, 36.25 grams for each), hence the name "Qianliang Cha" (tea weighing one thousand *liang*).

❶ 杀青 Fixing　❷ 揉捻 Rolling

千两茶制作技艺中以楠竹编织成篓，用棕叶铺篓，历经鲜叶杀青、揉捻、渥堆、松柴明火干燥、筛分、拼配、称重、蒸熟、装篓、自然晾晒干燥等23道工序，总工时近一年。茶叶经踩篓压制成形后，便可进行晒露。晒露，即将长约1.55米、直径约0.2米的圆柱体茶置于晾架上。经过夏秋季节的日晒、夜露、风吹（不能淋雨）之后，可进入长期存放阶段。在此过程中，茶叶在自然条件催化下自行发酵、干燥，并吸收蓼叶、篾片的香气，形成独特的风味。

❺ 制篓 Making bamboo baskets　❻ 装篓 Filling the basket

图3.76　千两茶
Qianliang Cha

When processing Qianliang Cha, it requires using baskets made of bamboo with palm leaves laid over the bottom of baskets. The processing usually takes up to one year and has 23 steps comprised of fixing, rolling, stacking, firewood drying, sieving, blending, weighing, steaming, picking, basket-filling and air-cooling and drying. Tea leaves will be sun-dried when stacked, compressed and shaped. For sun-drying, tea leaves are pressed into a cylinder some 1.55 meters long and 0.2 meters in diameter on a drying rack. Being exposed to the sun, dew and breeze (not rain) through summer and autumn, the cylinder will be stored for a long time. For this process, tea leaves ferment and dry on their

❸ 渥堆 Stacking

❹ 烘焙 Baking

❼ 踩篓 Basket piling

❽ 晒露 Sun-drying

图 3.77　千两茶制作工序
The production of Qianliang Cha

own under the influence of the natural environment. It will then absorb the aroma of the polygonum leaves and bamboo that gives Qianliang Cha a unique flavor.

制成的千两茶造型独特，汤色金黄如桐油，口感厚重，香气以松香为主，是中国茶中少有的一次性成形的单体重量较大的茶产品。

Processed Qianliang Cha has a unique shape, offering a golden brew like tung oil. It bears a thick taste and smells like rosin. Of all Chinese teas, Qianliang Cha is notable for its heavy weight.

3. 黑茶制作技艺（茯砖茶制作技艺）
Dark Tea Processing Techniques (Fuzhuan Tea)

黑茶制作技艺（茯砖茶制作技艺）主要流布于湖南省益阳市，于2008年列入第二批国家级非物质文化遗产代表性项目名录。

The processing of Fuzhuan Tea, prevalently seen in Yiyang, Hunan Province, was included in the second batch of national intangible cultural heritage representative list in 2008.

茯砖茶制作技艺较为复杂：先经毛茶采制、杀青、初揉、渥堆、复揉、干燥加工成黑毛茶，后经破碎、筛分、祛砂祛杂、除尘，拼配成清茶，再经炒茶、渥堆发酵、加茶汁、称重、汽蒸、散汽、关梆子等30多道工序制成。"发花"是茯砖茶制作技艺中特有的工序，可以使砖内产生名为"冠突散囊菌"的"金花"。

The production is complicated. The tea leaves must first be made into crude dark tea through picking, fixing, first rolling, stacking, re-rolling and drying, and then into Qingcha (pure tea) via crushing, sieving, cleaning, dedusting and blending. There are an additional 30 steps before Fuzhuan takes shape, including frying, stacking (for fermentation), tea juice adding, weighing, steaming, dissipating and modelling. The step "flowering" is particularly unique to Fuzhuan Tea, aimed to generate "eurotium cristatum", or "golden flowers", all over the tea bricks.

制成的茯砖茶，汤色橙黄明亮，香气纯正，滋味醇和，其中的茯茶素A和茯茶素B可促进人体新陈代谢。

The brewed Fuzhuan Tea is orange in color, giving a pure aroma and mellow taste. The lutein A and B it contains also help promote our metabolism.

图 3.78 茯砖茶
Fuzhuan Tea

图 3.79 金花
The golden flowers

广东
Guangdong

茶艺（潮州工夫茶艺）
The Tea Art (Chaozhou's Congou Tea)

茶艺（潮州工夫茶艺）主要流布于广东省潮州市，辐射周边地区，于2008年列入第二批国家级非物质文化遗产代表性项目名录。自明代朱元璋颁布"废团改散"诏令后，潮州地区逐渐流行瀹饮法散茶冲泡，至清代中期蔚然成风，冲泡方法已形成规范，并流传到东南亚等地。

Chaozhou's Congou Tea, so popular across Chaozhou and nearby areas in Guangdong Province, was included in the second batch of national intangible cultural heritage representative list in 2008. As Zhu Yuanzhang, the first emperor of the Ming Dynasty, issued an edict to "replace compressed cake tea with loose tea", the steeping method called "yue-yin" (loose tea brewing) took the central stage. It had become a common practice by the mid-Qing, when the brewing rules were established and spread to Southeast Asia.

潮州工夫茶艺中，多选用以凤凰单丛茶为代表的乌龙茶。冲泡程式有21道，包括"扇风催炭白""热盏巧滚杯""提铫速高注"，有"关公巡城池、韩信点兵准"的儒雅流畅，也有"先闻寻其香、再啜觅其味"的品饮之雅。

Oolong teas, like Fenghuang Dancong Tea, is ideal for making Congou Tea. The preparation of Congou Tea requires 21 steps, composed of burning charcoal, cup-warming, steeping while holding the pot high, etc. The procedures are usually named after legendary tales in history.

潮州工夫茶艺中三五成群共饮的习俗尤为普遍，蕴含了"和、敬、精、乐"的精神内涵。

The tea is usually served to people in small huddles, thus implying upon the philosophy of "harmony, respect, refinement and happiness".

图 3.80　淋壶
Watering teapot

图 3.81　"狮子滚绣球"
Cup rolling

图 3.82 "关公巡城"
"General Guan's Patrol"

十一

广西
Guangxi

1. 黑茶制作技艺（六堡茶制作技艺）
Dark Tea Processing Techniques (Liubao Tea)

黑茶制作技艺（六堡茶制作技艺）主要流布于广西壮族自治区苍梧县，于2014年列入第四批国家级非物质文化遗产代表性项目名录。

The processing of Liubao Tea, widely seen in Cangwu County, Guangxi Zhuang Autonomous Region, was included in the fourth batch of national intangible cultural heritage representative list in 2014.

六堡茶制作技艺分为杀青、揉捻、渥堆、初蒸、发酵、复蒸、干燥、晾置、加压、陈化等工序。

Its processing comprises fixing, rolling, stacking, primary steaming, fermentation, re-steaming, drying, cooling, compressing and aging.

图 3.83 鲜叶采摘
Picking fresh tea leaves

制成的六堡茶色泽黑褐光润，茶汤呈琥珀黄红之色，具有"红、浓、醇、陈"特色及突出的保健功效，被人们誉为"可以喝的古董"。

Liubao Tea appears glossy brown and yields an amber-like tea soup. It appears "red, thick with a mellow and aged taste" and claims health benefits. Therefore, it is called "a drinkable antique".

图 3.84 六堡茶
Liubao Tea

图 3.85 杀青
Fixing

图 3.86 揉捻
Rolling

2. 茶俗（瑶族油茶习俗）
Tea-Drinking Practices (Oil Tea of the Yao Ethnic Group)

茶俗（瑶族油茶习俗）主要流布于南岭走廊的瑶族聚居区，以广西壮族自治区桂林市恭城瑶族县为核心，辐射周边地区，于2021年列入第五批国家级非物质文化遗产代表性项目名录。

The custom of Oil Tea of the Yao ethnic group, prevalently seen in Yao ethnic community and its nearby areas, with Gongcheng Yao Autonomous County in Guilin of Guangxi Zhuang Autonomous Region being featured as the core area, was included in the fifth batch of national intangible cultural heritage representative list in 2021.

茶俗（瑶族油茶习俗）由古代的药引、粥茶、煎茶和煮茶衍变而来，油茶是瑶族人长期与山地潮湿的自然环境相处中形成的独特饮食习俗。

The Oil Tea of the Yao ethnic group, is derived from the past medicinal herbs, porridge tea, cooked tea and boiled tea. It is a very unique dietary practice of Yao people in their fight against the humid mountainous environment.

烹制油茶称为"打油茶"，以老叶绿茶为主料，配以生姜、猪油、盐等，放入铁锅捶打后煮熬，再加入米花、葱、蒜等。油茶味浓而涩，涩中带辣。有些地区还以脆果、花生、粑粑、汤圆等送服。瑶乡人打油茶，一般连做四锅，当地流传着"一杯苦、二杯涩、三杯四杯好油茶"的谚语。人们围坐在火塘边，边打油茶边聊家常。随着当地庙会的开展，油茶习俗得到了广泛传播，并逐渐传入壮族、汉族等人群，成为多民族共享的习俗。

The cooking process is termed as "da-you-cha" (the beating of oil tea). Aged green tea, being used as the base, will be cooked with a sprinkling of ginger, lard and salt. Then the mixture will be placed in an iron pot for continual "beating" and boiling, with the addition of pop rice, scallion and garlic. Oil tea tastes thick and a little spicy, and has a slight astringent effect. In some regions, crisp fruits, peanuts, cakes, or dumplings are served to go along with oil tea. When making it, Yao people usually steep it four times, as oil tea tastes much better on the fourth steeping. As the local saying goes, "Bitter first, then astringent, but on the third and fourth, the oil tea will turn good". It is usually done at parties where people chitchat around the campfire. Temple fairs also helped the oil tea be accepted to the rest of ethnic populations, like of Zhuang and Han nationality, making it now shared by multiple ethnic groups in China.

图 3.87　打油茶
Beating of Oil Tea

图 3.88 油茶
Oil Tea

十二

四川
Sichuan

1. 绿茶制作技艺（蒙山茶传统制作技艺）
Green Tea Processing Techniques (Traditional Processing Techniques of Mengshan Tea)

绿茶制作技艺（蒙山茶传统制作技艺）主要流布于四川省雅安市蒙顶山茶区一带，于2021年列入第五批国家级非物质文化遗产代表性项目名录。

Mount Mengding is home to the processing of Mengshan Tea. It was included in the fifth batch of national intangible cultural heritage representative list in 2021.

蒙山茶是多品类茶，包含蒙顶甘露、蒙顶黄芽、蒙顶石花、万春银叶和玉叶长春五种代表性茶品，以创制于明代的蒙顶甘露最负盛名。

The representatives of the Mengshan family are Mengding Ganlu, Mengding Huangya, Mengding Shihua, Wanchun Yinye and Yuye Changchun. Among them, Mengding Ganlu that originates from the Ming Dynasty is the best known one.

图 3.89　蒙山茶
Mengshan Tea

图3.90 摊凉
Cooling

蒙山茶传统制作技艺中,以制作蒙顶甘露为例,其核心技艺为红锅杀青、三炒三揉,炒制时在锅内抛撒翻炒,再进行推揉解块、做形提毫,最后烘焙提香。

In processing Mengshan Tea, say, the Mengding Ganlu, the core techniques involve high-temperature fixing and repeated frying and rolling for three times. Frying and flipping of tea leaves are done in a wok. Next comes the rolling and unblocking before having the leaves baked for a stronger fragrance, which is the final step of the entire process.

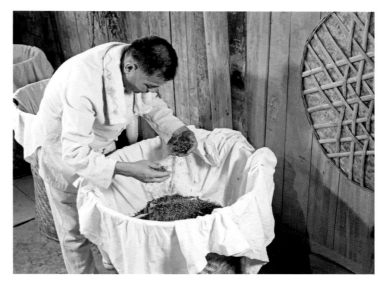

图3.91 烘焙提香
Baking for a stronger fragrance

2. 黑茶制作技艺（南路边茶制作技艺）
Dark Tea Processing Techniques (Nanlu Biancha)

黑茶制作技艺（南路边茶制作技艺）主要流布于四川省雅安市，于2008年列入第二批国家级非物质文化遗产代表性项目名录。

The processing of Nanlu Biancha, prevalently seen in Ya'an, Sichuan Province, was included in the second batch of national intangible cultural heritage representative list in 2008.

南路边茶制作技艺分为采割、原料茶初制、成品茶加工三部分，有重发酵、后发酵、多次发酵、非酶促发酵、转色发酵等特点。在初制中，有一炒、三蒸、三蹓（揉捻）、四渥堆、四晒、二捡梗、一筛分等18道主要工序；成品茶通过整理、拼配、压制等工序完成。

图 3.92　南路边茶
Nanlu Biancha

The processing involves three parts, namely the harvesting, pre-making and final processing. This is a sort of fermented tea, usually in heavy-fermented, post-fermented, multi-fermented, non-enzymatic fermented and color-converting fermented categories. There are, in pre-making, 18 major steps covering stir-frying (one time), steaming (three times), rolling (three

图3.93 蹓茶
Foot-rolling

times), stacking (four times), drying (four times), stalk-picking (twice) and sieving (one time). The final processing comprises sorting, mixing and compressing.

制成的南路边茶外观褐黑油润，汤色褐红明亮，香气浓郁持久，滋味醇和悠长，加入酥油、盐、核桃仁末等搅拌可制成酥油茶。

Nanlu Biancha has a brownish-black and satiny appearance. The brew has a lasting aroma and a mellow, lingering taste. If mixed with yak butter, salt and ground walnuts, Nanlu Biancha can be made a sort of buttered tea.

十三

贵州 Guizhou

绿茶制作技艺（都匀毛尖茶制作技艺）
Green Tea Processing Techniques (Duyun Maojian)

绿茶制作技艺（都匀毛尖茶制作技艺）主要流布于贵州省都匀市，于2014年列入第四批国家级非物质文化遗产代表性项目名录。

The processing of Duyun Maojian, popularly seen in Duyun, Guizhou Province, was included in the fourth batch of national intangible cultural heritage representative list in 2014.

都匀毛尖茶制作技艺包括采摘、杀青、揉捻、搓团、提毫、烘焙等工序。整个过程中，抛炒、抖闷、抖揉、揉捻的方向、次数及用力轻重等随茶青的质量、湿度，以及加工时间、程序、锅温的不同而有所差异。

Its processing covers steps like picking, fixing, rolling, rubbing, trichome-exposing (ti-hao) and baking, etc. Processing parameters, like the kneading direction, quantity used or the strength used for tossing, frying, shaking and rolling, can vary with the leaves' quality, humidity, processing time and the baking temperatures.

制成的都匀毛尖茶外形条索卷曲，色泽鲜绿，白毫显露，汤色清澈，滋味鲜浓，叶底淡黄明亮。

The curly-shaped Duyun Maojian looks shinning green with white trichome-exposing. It produces a clean brew that is fresh and thick in flavor. Its post-brewing leaves has a bright, pale yellow shade.

图 3.94　都匀毛尖茶
Duyun Maojian

图3.95 控制锅温
Control the wok temperature

图3.96 搓团提毫
Rubbing and trichome-exposing

图 3.97 贡茶
Tribute Tea

十四

云南 Yunnan

1. 普洱茶制作技艺（贡茶制作技艺）
Pu'er Tea Processing Techniques (the Tribute Tea)

普洱茶制作技艺（贡茶制作技艺）主要流布于云南省宁洱哈尼族彝族自治县，于2008年列入第二批国家级非物质文化遗产代表性项目名录。

The Pu'er Tea processing (the Tribute Tea), prevalently adopted in Ning'er Hani and Yi Autonomous County, Yunnan province, was included in the second batch of national intangible cultural heritage representative list in 2008.

贡茶制作技艺，包含祭祀茶神、原料采选、杀青揉晒、蒸压成形四个程序。茶叶采摘开始前向茶神行敬献仪式。仪式结束后，按一定标准采选原料。然后进入杀青揉晒环节，以特定工艺将鲜叶加工成晒青茶。随后通过蒸、揉、压、定形、干燥、包装等工序将晒青茶蒸压成形，制成各种成品茶。

图3.98 揉捻
Rolling

The procedures of making Tribute Tea comprise tea-god worship, leaves picking, fixing (with rolling and drying) and steaming for shaping. Before picking tea leaves, a sacrificial ceremony must be performed in honor of the god of tea. After the ceremony, there are certain regulations on the picking of the leaves. It will be followed by the fixing (with rolling and drying) part, which is to convert fresh leaves into sun-dried tea. The sun-dried tea is the base of final teas through steaming, rolling, pressing, shaping, drying and packaging.

2. 普洱茶制作技艺（大益茶制作技艺）
Pu'er Tea Processing Techniques (Dayi Tea)

普洱茶制作技艺（大益茶制作技艺）主要流布于云南省勐海县一带，于2008年列入第二批国家级非物质文化遗产代表性项目名录。

The processing of Dayi Tea, prevalently seen in Menghai, Yunnan province, was included in the second batch of national intangible cultural heritage representative list in 2008.

大益茶制作技艺中，以云南大叶种茶为原料，先制成晒青毛茶，再根据不同需求制成茶品。整个技艺包括30多道工序，即毛茶加工、发酵、筛分拣剔、拼配、成形、包装等。技艺的关键在于拼配和发酵。拼配，即根据茶叶品种特点进行组合，弥补单一茶青的不足；发酵，则特指"人工后发酵"，即渥堆。

The production of Dayi Tea uses the big-leaf tea specimens as raw materials. First have the leaves sun-dried, then processed into various teas according to specific requirements. The entire production comprises more than 30 steps, including base processing, fermentation, sieving and screening, blending, molding and packaging. Blending and fermentation are the more crucial steps. Teas will be blended according to their properties. This will help make up for the shortcomings of using a single tea type. Fermentation refers to a special "artificial post-fermenting" technique, or to put it simply, stacking.

图3.99 晒青毛茶
The leaves sun-dried

图 3.100 干燥
Drying

3. 黑茶制作技艺（下关沱茶制作技艺）
Dark Tea Processing Techniques (Xiaguan Tuocha)

黑茶制作技艺（下关沱茶制作技艺）主要流布于云南省大理白族自治州，于2011年列入第三批国家级非物质文化遗产代表性项目名录。下关沱茶因产于大理的下关镇而得名。

The processing of Xiaguan Tuocha, widely found in Dali Bai Autonomous Prefecture, Yunnan Province, was included in the third batch of national intangible cultural heritage representative list in 2011. Xiaguan Tuocha was named after its birthplace Xiaguan Town in Dali.

下关沱茶制作技艺中，以云南大叶种晒青茶为原料，技艺主要为筛分、拣剔、拼配、称量、蒸揉、压制成形、干燥、包装等。其中压制成形是技术关键，采用有凹凸槽模的木凳，以杠杆原理进行压制，之后在布袋内冷却，等定形后解开布袋，置于木框上晾干。整个过程采用"细茶精制，粗茶细制，精提净取"原则。

The production of Xiaguan Tuocha uses Yunnan's big-leaf green tea as its base, and the entire process comprises sieving, screening, blending, weighing, steaming and rolling, compressing, drying, and packaging. Of the above steps mentioned, compressing is key to the final success. The grooved wooden bench is designed to be able to compress tea leaves, and they are then placed in cloth bags for cooling. After the cooled leaves have hardened, they will be placed on a wooden frame for drying. The principle of refining the tea whether fine or otherwise will be adhered to during the entire process.

图 3.101 下关沱茶
Xiaguan Tuocha

图 3.102 称茶
Weighing

图 3.103 揉茶
Rolling

图 3.104 压茶
Compressing

图 3.105　萎凋
Withering

4. 红茶制作技艺（滇红茶制作技艺）
Black Tea Processing Techniques (Dianhong Tea)

红茶制作技艺（滇红茶制作技艺）主要流布于云南省凤庆县，于2014年列入第四批国家级非物质文化遗产代表性项目名录。

The processing of Dianhong Tea, prevalently seen in Fengqing County, Yunnan Province, was included in the fourth batch of national intangible cultural heritage representative list in 2014.

滇红茶制作技艺的初制工序主要为萎凋、揉捻、发酵、干燥。有"初制把五关"之说，即鲜叶选取关、萎凋时效关、揉捻时效关、发酵时效关、干燥火候关。萎凋可适当蒸发水分，使梗叶变成萎蔫状态。揉捻可以破坏叶细胞，增进茶叶的色、香、味浓度。发酵使叶细胞在酶的作用下进行氧化反应，叶色由绿变为微红或菜花黄，青草气消失，出现熟果香，是形成滇红茶"红叶红汤"的重要工序。干燥则采用高温烘焙，最终获得滇红茶特有的甜香。

The primary processing of Dianhong Tea includes withering, rolling, fermenting and drying, requiring stringent control in five aspects, namely the selection of fresh leaves, the time period required for withering and rolling, fermentation, and the time duration and degree of drying. The part of withering can get water evaporated properly and stems and leaves wilted. Rolling can destroy the leaf cells and thus enhance the color, aroma and flavor of the tea brew. Fermenting allows the leaf cells undergo oxidation reactions thanks to enzymes, so that green leaves will turn slightly reddish or cauliflower yellow, and the "smell of grass" will give way to the fragrance like ripe fruits. This is an important technique to assure the "red brew" of Dianhong Tea. The drying, done through high temperature baking, will give Dianhong Tea a unique sweet aroma.

制成的滇红茶汤色呈棕色、粉红或姜黄，颜色鲜亮，在茶汤中加入适量牛奶仍有较浓茶味。

The steeped Dianhong Tea usually assumes a clear brown, pink or turmeric color. Its strong flavor will be retained even with milk.

图 3.106　发酵
Fermentation

图 3.107　干燥
Drying

5. 茶俗（白族三道茶）
Tea-Drinking Practices (Three-Course Tea of the Bai Ethnic Group)

茶俗（白族三道茶）主要流布于云南省大理白族自治州，于2014年列入第四批国家级非物质文化遗产代表性项目名录。三道茶是白族一种古老的习俗，起源于唐代南诏国时期，演变成白族民间礼宾待客的传统礼俗。这是一种宾主抒发感情、寄托美好祝愿并富于戏剧色彩的饮茶方式。

The Three-Course Tea of the Bai ethnic group, popular in Dali Bai Autonomous Prefecture, Yunnan Province, was included in the fourth batch of national intangible cultural heritage representative list in 2014. It originates from the period of Nanzhao Kingdom and has now been accepted as a folk tradition of the Bai people to receive guests. That's the way to express their feelings and make wishes.

白族三道茶的制作非常考究，烤茶注重烤器和火候，冲泡时对水温有一定要求，其最具特色的就是分"三道"饮用，即"一苦、二甜、三回味"的饮茶方式。三道茶第一道为"苦茶"，又因开水入罐时发出沸腾声而得名"雷响茶"，以苍山绿茶为主，味道甘苦；第二道为"糖茶"或"甜茶"，主材料为红糖、乳扇、核桃仁片，寓意苦尽甘来；第三道为"回味茶"，主材料为蜂蜜、花椒、桂皮，味甜微麻又略苦，寓意对一生的回味。三道茶，三种口味，先苦后甜，意义深远。

The processing of the Three-Course Tea is sophisticated. It requires proper utensils and fire control in baking, and water temperature in steeping. Like the name, it actually consists of three cups of tea. The first one tastes "bitter", the second one "sweet" and the third "with a lingering aftertaste". The first course is "bitter tea", which is made basically of Cangshan Green Tea that tastes refreshingly bitter. It is

图 3.108 白族三道茶
Three-Course Tea of the Bai ethnic group

also known as "thunder tea" as there will be a thunder-like sound when boiling water is being poured in. The second course, termed "sugar tea" or "sweet tea", is prepared with brown sugar, dairy fan and walnut slices, signifying a reward for the past tough moments. The third course is named "aftertaste tea". It is made of honey, pepper and cinnamon. The tea is sweet, yet albeit a spicy and bitter flavor, implying a reflection of one's life. The three courses have three flavors that, as first bitter and then sweet, gives a thought-provoking message.

图 3.109　三道茶原料
Ingredients for Three-Course Tea

图 3.110　制作三道茶
Preparing the Three-Course Tea

6. 德昂族酸茶制作技艺
Sour Tea Processing Techniques of the De'ang Ethnic Group

德昂族酸茶制作技艺主要流布于云南省芒市，于2021年列入第五批国家级非物质文化遗产代表性项目名录。德昂族是芒市的世居民族之一，有"古老茶农"的美誉。德昂族酸茶分为湿茶和干茶，湿茶食用，干茶饮用。

The Sour Tea processing of the De'ang ethnic group, mostly in Mangshi, Yunnan Province, was included in the fifth batch of national intangible cultural heritage representative list in 2021. De'ang ethnic group is one of the longest-dwelling people of Mangshi and they are known as "the ancient tea farmers". The Sour Tea can be divided into wet tea and dry tea, with the former being a dish and the latter a beverage.

德昂族酸茶制作技艺中，采用云南大叶种茶叶为原料，工序分为杀青、揉捻、厌氧发酵、舂制（捣碎）、做形、干燥等。厌氧发酵是德昂族酸茶制作技艺的关键，也是形成酸茶品质特征的重要环节。通常是将揉捻叶放入竹筒或土罐内压紧压实，再将密封后的竹筒口朝下埋于地下，或将罐子密封后置于近似地下温湿度的地方，在微生物、酶和湿热的共同作用下进行厌氧发酵。食用茶的发酵时间约为2个月，饮用茶的发酵时间为4~9个月。酸茶独特的发酵工艺，使得茶叶产生多种益生菌。

The Sour-Tea making requires fresh leaves of Yunnan's big-leaf tea species as the base, and entire procedures comprise fixing, rolling, anaerobic fermentation, pounding (or mashing), shaping and drying. Anaerobic fermentation is key to De'ang Sour Tea processing and a critical step that determines tea quality. Tea leaves, as being rolled, are put into bamboo tubes or earthen jars and compacted. Bury the sealed tubes into the ground with their mouth facing downward, or place the sealed jars in a place with temperature and humidity close to those of the underground. Anaerobic fermentation will be taking effect under the joint actions of microorganisms, enzymes and moist heat. It usually takes 2 months to ferment to dish and 4–9 months to make it a beverage. The processing of Sour Tea helps produce a range of probiotic bacteria.

完成发酵后，或将茶叶舂制成茶泥，制成茶饼，经干燥后切割为片茶；或直接干燥为散茶。

It will finally come in two forms, namely either tea cakes as fermented that can be made slices or loose tea.

制成的德昂族酸茶因年限不同，汤色可呈现出黄绿、金黄透亮或红色等不同颜色，嗅之微酸，饮之轻柔爽口，既有熟茶的柔和，又不失绿茶的清新，回味甘甜。

The Sour Tea's brew, based on different ages, appear in various hues, like yellow-green, gold or pure-red. With a slight sour smell, and yet a soft and refreshing taste, the sour tea still retains a clear and sweet aftertaste of green tea.

图 3.111 德昂族酸茶
Sour Tea of the De'ang ethnic group

❶ 杀青 Fixing

❷ 蒸茶 Steaming

❸ 厌氧发酵 Anaerobic fermentation

❹ 春制 Pounding

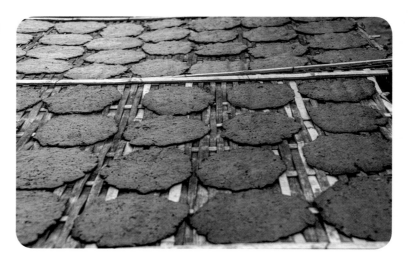

❺ 干燥 Drying

图 3.112 德昂族酸茶制作工序
The production of Sour Tea of the De'ang ethnic group

十五

陕西
Shaanxi

黑茶制作技艺（咸阳茯茶制作技艺）
Dark Tea Processing Techniques (Xianyang Fucha)

黑茶制作技艺（咸阳茯茶制作技艺）主要流布于陕西省咸阳市一带，于2021年列入第五批国家级非物质文化遗产代表性项目名录。咸阳茯茶有"自古岭北不植茶，唯有泾阳出砖茶"之说。

The processing of Xianyang Fucha, prevalently seen in Xianyang, Shaanxi Province, was included in the fifth batch of national intangible cultural heritage representative list in 2021. There has been a saying about it that "Although the North of the Five Ridges is not a place to grow teas, Jingyang (a county belonging to Xianyang) is an exception".

咸阳茯茶制作技艺，一般要经过30多道工序，其中"发花"是关键环节。经过发酵自然生成的"金花"是形成茯茶独特品质的关键，与茶叶的滋味、香气密切相关。

The making of Fucha comprises more than 30 steps, of which the "flowering" one is the most crucial. It is the "golden flowers" naturally born out of fermentation that will determine the property of the tea, just like its aroma and taste.

作为历朝各代用于"茶马交易"的主要茶品，咸阳茯茶正焕发出勃勃生机。

Xianyang Fucha, as the major commodity along "the Old Tea-Horse Road" in the past, is still alive and even now getting more popular in China.

图 3.113　包装
Packaging

图 3.114　咸阳茯茶
Xianyang Fucha

第三章 | 寻茶问水 咀华啜英

Tea Drinking Practices

藏族酥油茶
Tibetan Buttered Tea

酥油茶，藏语俗称"甲脉儿"，藏族民间传统日常饮料，流行于西藏、青海、甘肃、四川及云南等藏族人民居住地区。其制法为先将砖茶用水久熬成红色浓汁，倒进酥油茶桶（藏语称"董莫"）内，放入酥油和食盐，用力将"甲罗"（酥油茶桶内的木质搅拌器）上下来回抽几十下，使其水乳交融即成。酥油茶是藏族人民日常生活中的主食之一，亦是待客的必备饮料。喝酥油茶有一套规矩，主人请喝酥油茶，一般是边喝边添，不一口喝完；如喝了一半不想再喝，待主人再次添满后，或客人在辞行时应一饮而尽，这样才符合藏族的习惯和礼貌。

Known as "Jiamaier" in Tibetan, Buttered Tea is a traditional Tibetan folk daily drink, and popular in Tibetan areas such as Tibet, Qinghai, Gansu, Sichuan and Yunnan. The making method is: First, boil the brick tea with water for a long time into red thick juice, which is then poured into the buttered tea bucket (called "Dongmo" in Tibetan), butter and salt added. Second, forcibly stir with "Jialuo" (a wooden stirrer inside the bucket) up and down for dozens of times to blend those things well. Buttered Tea is a staple food for the Tibetans, and also a must drink for guests. There is a set of rules to follow in Buttered Tea drinking. One can't finish the tea at one go. Only can one do so when he takes leave and the host fills the cup with tea for the last time.

藏族居住地区多干旱寒冷，以奶、肉、糌粑等为主食，缺少蔬菜瓜果等物。茶中富含咖啡碱、茶多酚、维生素，具有清热、润燥、解毒、利尿等功能，能弥补藏族饮食中的缺陷，是藏族人民不可缺少的日常食品。

Much of the Tibetan areas are dry and cold, with milk, meat, tsamba as staple foods, lacking vegetables, fruits, etc. Rich in caffeine, tea polyphenols and vitamins, the Buttered Tea can reduce heat and dryness, detoxify the body, induce urination, etc., thus making up for the defects in Tibetan diet and becoming an indispensable drink for Tibetan people.

扫码了解藏族酥油茶

图 3.115　藏族室内饮茶场景
Tea drinking of the Tibetans (indoors)

二

维吾尔族奶茶与香茶
Uygur Milk Tea and Fragrant Tea

饮茶在维吾尔族人的生活中占有重要的地位，民间有"宁可一日无米，不可一日无茶""一日三餐有茶，提神清心，劳动有劲，三天无茶落肚，浑身乏力，懒得起床"的说法。维吾尔族主要分布在新疆维吾尔自治区，由于新疆地域辽阔，天山南北气候各异，饮茶习俗也有所不同。除奶茶和香茶外，维吾尔族还有果仁茶、药茶、炒面茶等多种茶饮。无论是制作奶茶还是香茶，维吾尔族多用茯砖。而随着时代的发展，如今的维吾尔族人民饮用的茶类也趋于多样化，红茶、绿茶、花茶等茶类也逐渐受到欢迎。

Tea drinking plays an important role in the life of Uygur people, as shown in some folk sayings: "It is better to have no rice a day than no tea a day"; "Having tea for three meals a day refreshes your mind and makes you work vigorously. Having no tea for three days makes you tired and reluctant to get up". The Uygur people are mainly living in Xinjiang Uygur Autonomous Region. The vastness of Xinjiang, and climatic differences between the north and south of Mount Tianshan, has generated different tea drinking customs there. In addition to milk tea and fragrant tea, Uygur people also consume nut tea, medicinal tea, fried flour tea and other drinks. Fuzhuan Tea is widely used to make milk tea or fragrant tea. As time goes by, Uygur people today drink a variety of teas, including black tea, green tea and scented tea.

1. 南疆香茶
Fragrant Tea in South Xinjiang

南疆地区以农耕为主，当地的维吾尔族多喜香茶。香茶在煮茶的过程中加入了胡椒、桂皮等香料。维吾尔族喝香茶，习惯于一日三次，与早、中、晚三餐同时进行，通常是一边吃馕，一边喝茶。南疆维吾尔族的茶具常见为长嘴铜壶和瓷茶碗。

图 3.116　维吾尔族室内饮茶场景
Tea drinking of the Uygur people (indoors)

South Xinjiang relies heavily on farming, and the local Uygur people prefer fragrant tea, which is a tea boiled together with pepper, cinnamon and other spices. Uygur people drink fragrant tea three times a day at breakfast, lunch and dinner, usually drinking tea while eating naan. Brass long-spout teapots and porcelain tea bowls are commonly used among the Uygur people in South Xinjiang.

2. 北疆奶茶
Milk Tea in North Xinjiang

北疆以畜牧业为主，当地维吾尔族牧民多饮奶茶。奶茶制作通常是在煮好茶汤后加入鲜奶或奶皮子，滚沸时不断用勺扬茶，直到茶乳充分交融，最后加盐调味。新疆地区的其他民族如蒙古族、哈萨克族等都有饮用奶茶的习俗。

North Xinjiang is dominated by animal husbandry. The local Uygur herdsmen like drinking milk tea. It is prepared by boiling tea and then adding fresh milk or vrum in it. During the boiling, one should constantly stir the tea with a spoon until the tea and milk are completely mixed. Finally salt should be added. Mongolian, Kazakh and other ethnic groups in Xinjiang drink milk tea too.

三

徽商茶庄
Anhui Merchants' Tea Shop

徽州商人"其货无所不居"，经营范围极广，而茶业是其支柱行业之一。徽州自古以来就是我国重要的产茶区之一。南唐刘津曾说："婺源、浮梁、祁门、德兴四县茶货实多。"唐张途《祁门县新修阊门溪记》中载："邑（祁门）山多而田少，水清而地沃。山且植茗，高下无遗土，千里之内，业于茶者七八矣。由是给衣食，供赋役，悉恃此。"垂至明清，随着茶叶需求的猛增，徽州茶叶的生产有了长足的发展，一大批徽州茶商应运而生，而他们的商业网络，几乎覆盖了大半个中国。可以说徽商的茶业经营活动在我国明清社会经济领域中发挥了重要作

PART IV

WORLDWIDE SPREAD OF TEA

第三章

世界茶业
Tea Industry in the World

一 世界茶业大事记 A Chronicle of Events of the Global Tea Business …… 366

二 世界茶业现状 The Global Tea Business Today …… 373

用，它的兴衰与当时的社会经济条件息息相关。

Huizhou merchants engaged in a wide range of business, and tea trade is one of their pillar industries. Traditionally Huizhou was one of the important tea producing areas in China. Liu Jin of the Southern Tang Dynasty said, "The four counties of Wuyuan, Fuliang, Qimen and Dexing produce teas in large quantity." Zhang Tu, a Tang official, recorded in his account: "The county (Qimen) is mountainous and it suffers from a shortage of arable land, while the land is fertile. Tea is planted almost everywhere in the mountains. Within a thousand miles, 70 to 80 percent of the population involves in tea industry, which serves a resource for their food, clothing, taxes and duties." In Ming and Qing Dynasties, with the soaring demand for tea, Huizhou tea production made substantial progress, giving birth to a large number of Huizhou tea merchants, whose business network covered more than half of China. As it were, Huizhou merchants and their tea business played a crucial role in the social and economic fields of the Ming-Qing Dynasties, and its rise and fall were closely linked with the social and economic conditions at that time.

四
川蜀茶馆
Sichuan Tea House

一城居民半茶客。川蜀茶馆是中国饮茶风俗及茶馆文化非常有特色的代表。在川蜀地区，茶馆也称为茶铺，它不仅是人们休闲小憩，摆"龙门阵"的去处，旧时也是吃茶讲理、解决纠纷、消释前嫌的断案场所，是当地居民重要的公共空间。

City residents, teahouse regulars. Sichuan tea house is very typical of Chinese tea-drinking customs and tea house culture. Also referred to as "chapu" (tea shop) there, the tea house is a place where people take a rest and chat for hours. In the past, the local residents would settle disputes and dispel grievances over cups of tea in the tea house, the significant public space.

晚清知县周询曾提到："茶社无街无之，然俱当街设桌，每桌四方各置板凳一，无雅座，无楼房，且无倚凳，故官绅中无人饮者。"20世纪初，川蜀一带的茶馆开始使用矮方木桌和竹椅。

As described by Zhou Xun, a county magistrate of the late Qing Dynasty, "There are tea houses in every single street. Square tables, with four benches, are placed in the open air. No private rooms, no houses, and no leaning stools, so none of officials or gentlemen will go and drink tea there." At the beginning of the 20th century, low square wooden tables and bamboo chairs were used in Sichuan tea houses.

川蜀茶馆多用铜茶壶、锡杯托、瓷盖碗、沱茶或花茶。茶馆中茶博士的掺茶技巧堪称一绝，也是川蜀茶馆中一道独特的风景。

Copper teapots, tin cup holders, porcelain covered bowls, tuocha or scented tea are usually used in Sichuan tea houses. Waiters in the tea house have fabulous skills in tea serving, which is also a unique scene in Sichuan tea houses.

扫码了解川蜀茶俗

茶传

第四篇

第一章

芳茶远播
Spread of Tea and Tea Culture

315

一　万里茶路　The Long Long Tea Road …… 315

二　茶音传承　Tea Pronunciation Spread …… 319

第二章

五洲茶话
Tea Culture of the Five Continents

332

一　亚洲茶文化　Asian Tea Culture …… 332
二　欧洲茶文化　European Tea Culture …… 344
三　非洲茶文化　African Tea Culture …… 352
四　美洲茶文化　American Tea Culture …… 358
五　大洋洲茶文化　Oceania Tea Culture …… 363

第一章 芳茶远播

Spread of Tea and Tea Culture

茶叶植根于中国大地，枝繁叶茂，葱翠可喜。千百年来，历史的机缘使其走出国门，远播海外，同行的还有中国的茶叶加工工艺、植茶技术、饮茶方法、茶事礼俗、精美茶具等。世界各地的茶叶，均直接或间接来自中国。茶叶之路，亦是中国文化的传播之路。茶风所至，许多事物悄然改变。

Tea grows in China originally. In thousands of years, tea left China and spread abroad, so did Chinese techniques of tea processing and planting, drinking methods, tea-related etiquettes and customs, exquisite tea sets and so on. Tea in the other countries came directly or indirectly from China. Where there was tea, there was Chinese culture. Much has changed quietly because of Chinese tea.

一 万里茶路
The Long Long Tea Road

自从人类发现茶，认识茶的妙处之后，茶便为人们所青睐。与此同时，饮茶习俗、茶树栽培与加工技术等逐渐开始扩散。中国茶文化向国外传播，始于唐代，盛于明清，并在现代进一步发展。其对外传播的途径主要有两条路线，即陆路传播与海路传播。

Tea has become a much-favored beverage since its beauties were unveiled. Meanwhile, customs to drink and techniques to grow and process tea were spreading worldwide. It was in the Tang Dynasty (618–907) that the Chinese tea culture was first known to the world. Such cultural communications went viral over the Ming and Qing Dynasties (1368–1911) and saw its peak in modern years. Land and sea are both important ways to channel the culture overseas.

1. 陆路传播 Spread by Land

茶叶在10—12世纪时，已经由吐蕃传到高昌、于阗和七河地区。进入13世纪，中西陆海交通打开，茶进一步在中亚和西亚传播。

Between the 10th and 12th centuries, tea came from Pugyel to Khoco, Khotan and Zhetysu. In the 13th century, the overland and marine traffic between China and the West were opened, and tea was further spread in Central and West Asia.

图4.1 茶马古道遗址
The site of the Old Tea-Horse Road

2. 海路传播 Spread by Sea

茶叶、丝绸、瓷器等是中国海上贸易的重要商品。7世纪，茶叶向东传往朝鲜半岛和日本。13世纪之后，茶叶销往东南亚。17世纪之后，航海技术的发展促使东西方贸易往来频繁，茶叶也传往更遥远的欧洲、美洲、非洲等地。

Tea, silk and porcelain are important goods of China's maritime trade. In the 7th century, tea spread eastward to the Korean Peninsula and Japan. After the 13th century, tea was sold to Southeast Asia. After the 17th century, the frequent trade between the East and the West was promoted by the development of navigation technology, and tea also went to more distant places such as Europe, America and Africa.

图4.2 茶叶之路上的东正教教堂
Orthodox church on the Tea Road

扫码了解华茶传播

图4.3 运输茶叶的驼队
Camels carrying tea

图4.4 中荷茶叶贸易
Sino-Dutch tea trade

图4.5 19世纪末中国茶叶大量外销

Chinese tea was exported in large quantities at the end of the 19th century

二

茶音传承
Tea Pronunciation Spread

目前，世界各国对"茶"字的读音，都和中国茶的传播和出口地区人们的读音相近。大致说来，可分为两大读音体系：一是中国普通话语音"茶"——"Cha"；一是中国福建厦门地方语音"的"——"Tey"。两种语音，在对外传播时间上有先有后，路径亦不尽相同，与中国茶向外传播的路线基本一致。

The pronunciation of the word "tea" is similar to that in the regions where Chinese tea is exported. It can be roughly divided into two pronunciation systems: Chinese Putonghua pronunciation "Cha"; The local pronunciation of "Tey" in Xiamen, Fujian, China. The two sounds spread at different time and by different means, which are basically the same as the spread route of Chinese tea.

1. "Cha"

一般说来，"Cha"的发音首先主要传播至中国的邻国，这些国家由于临近中国，最早接触到茶。日语、印度语、巴基斯坦语、孟加拉国语、波斯语（伊朗、阿富汗）、土耳其语、俄语等中的"茶"，基本为"茶"字原读音。此外，在乌尔都语、奥利亚语、朋巴拉语等语言中，"茶"字的读音也都源于汉字"茶"的读音。

Generally speaking, pronunciation of "Cha" first spread to China's neighboring countries. Because of their close proximity to China, they were the first to come into contact with tea. For example, the word "tea" in Japanese, Hindi, Pakistani, Bengali, Persian (Iran, Afghanistan), Turkish, Russian, etc. is basically sounded like the Chinese "Cha". In addition, in Urdu, Oriya, Bambara, etc. the pronunciation of the word "tea" is also derived from the Chinese way of saying "Cha".

2. "Tea"

明末清初，一些西方远洋航行船队在运输茶的同时，也吸收了"茶"在中国当地方言中的发音。英语、法语、德语、荷兰语、拉丁语、西班牙语等对茶的称谓，均受福建、广东沿海地区方言的影响。

In the late Ming and early Qing Dynasties, when some western oceangoing fleets transported tea, they borrowed the sound of "tea" from the local Chinese dialects. How to say "tea" in English, French, German, Dutch, Latin and Spanish languages was influenced by the dialects of the coastal areas of Fujian and Guangdong.

晚清 龙纹竹节银茶具

茶壶（中）：高 14、口径 8、底径 8 厘米

奶杯（左）：高 7.8、口长 10、底径 5.7 厘米

糖缸（右）：高 11、口径 6.5、底径 6.3 厘米

  由茶壶、奶杯和糖缸组成。茶壶壶纽、把手及流均制成竹节状，壶身锤鍱錾刻一条飞龙，环绕壶身，居中为一素面盾形徽章，底承覆扣喇叭形圈足。壶把手上下端均嵌有象牙，可以起到隔热的作用。壶盖与壶身以合页连接为一体。奶杯把手为竹节状，糖缸盖纽、两耳为竹节状，其余器身纹饰与茶壶相同。

  该套茶具为典型的西方茶具形制，但纹饰采用了中国传统的竹节纹和龙纹，反映了当时外销银器中西交融的独特风貌。

茶杯

清 青花诗文茶具

茶杯：高 5.5、口径 7.4、底径 3.3 厘米

茶托：高 3、口径 18.6、底径 13.3 厘米

茶壶：高 17.5、口径 7.8、底径 12.5 厘米

  由茶杯、茶托及茶壶组成。

  杯敞口，深腹，圈足。托敞口，深腹，折腰，矮圈足。杯腹和茶托内底以青花书楷书七言诗句，摘自中国唐宋诗文，青花发色鲜艳，字体端正。杯底和托底均以青花书"满堂福记"四字双行行书款。

  壶直口，短颈，折肩，圆筒形腹，圈足，内嵌式折沿盖，三弯圆流，高提梁以铜制成。壶盖、肩、腹以及流均以青花书《朱夫子家训》，家训以楷书写成。铜制提梁系瓷器外销泰国后添加。

  清代中期出现大量带有中国传统文化标识的外销茶具，深受国外消费者的喜爱。

茶托

南方嘉木：中华茶文化

第四篇 茶传五洲

茶壶

清 锡胎包椰壳雕杂宝纹茶具

软提梁壶（左）：高 10、口径 6、底径 6.5 厘米
奶杯（右）：高 11、口长 7、底径 4.6 厘米
茶叶罐（中）：高 15.3、口径 5、底径 7.5 厘米

由软提梁壶、奶杯及茶叶罐组成。

软提梁壶内凹口，盖纽制成瓜蒂状，弧肩，扁圆形腹，矮圈足，弧腹一侧置一三弯形流，肩部两侧立两竖耳，穿孔，有两环形提梁。奶杯是西方奶茶具的主要器物之一，口沿设计巧妙，一侧安一曲形把，器腹呈长圆形。茶叶罐直口带盖，圆鼓腹，撇圈足。

三件器物均以金属锡为内胎，捶打成形，提梁壶和奶杯的下腹部及底部、茶叶罐的腹部包以椰壳作为装饰，椰壳周身雕刻杂宝纹和团寿纹。

明清两代，以锡为材料制作茶具非常普遍，也备受推崇。一方面，锡的延展性较好，很容易塑形，可根据制作者或消费者需要制作造型各异的茶具；另一方面，锡可与紫砂、玉、牛角等各种其他材料结合，相得益彰。

328
329

第二章 五洲茶话
Tea Culture of the Five Continents

中国的茶叶漂洋过海，走进异国他乡后，便逐渐融入异域文化，在异国生根、繁衍、蔓延，开出似锦繁花。世界茶文化的篇章，由中国书写在前，笔墨酣畅淋漓；其他国家与地区提笔在后，风格独树一帜。各种涓滴细流，最终汇成茶文化的巨川，奔流不息。

After traveling across the sea and entering foreign countries, Chinese tea has gradually blended with foreign cultures, took root, spread and blossomed in foreign countries. The tea culture was created first by China, and then by other countries and regions. All sorts of trickles eventually converge into a huge stream of tea culture.

一
亚洲茶文化
Asian Tea Culture

在广袤的亚洲大陆，虽然地域分东西南北中，人们却拥有共同的语言：茶。以茶消乏解渴，以茶助兴佐谈，以茶言志表情，以茶入诗入画，茶在亚洲常有不一般的地位。茶中有"艺"，茶中有"礼"，发展到极致时，茶中有"道"，有说不尽的意蕴。

In the vast Asian continent, although people are living in different places, they share a common language: tea, which is used to eliminate fatigue and quench thirst, to entertain and stimulate conversation, to express aspirations and feelings, and to be committed to paper. There is "art", "courtesy" and even "Tao" in tea.

扫码了解亚洲茶文化

1. 日本 Japan

唐宋时期，茶籽、饮茶方法等传入日本，在此过程中发挥了重要作用的有日本僧人最澄、荣西、圆尔等。例如，1241年，留学僧圆尔从浙江径山带回《禅苑清规》、径山茶种和饮茶方法，并制定寺院清规，将茶礼列为禅僧日常生活中必须遵守的行仪作法。1259年，南浦绍明将径山茶宴系统地传入日本。之后，经过村田珠光、武野绍鸥和千利休等人的完善，日本最终形成了源自中国而又别具特色的茶道。

Tea seeds, together with methods of tea preparation, were introduced to Japan in the Tang and Song Dynasties. The Japanese monks contributing to this include Saichō, Eisai and Enn Ni. For example, in 1241, Enn Ni, who studied in the Southern Song, brought back from Jingshan, Zhejiang, *Pure Rules for Chan Monasteries* (Chanyuan Qinggui), Jingshan tea seeds and tea drinking methods. Based on the text, he formulated monastic codes, and listed the tea ceremony as the ritual practice that the monks must observe in daily life. In 1259, Nanpo Syoumyou introduced systematically the Jingshan Tea Banquet into Japan. After that, Murata Syukou, Takeno Zyou Ou, Sen no Rikyū and others strove to better it, finally leading to unique Japanese tea ceremony of Chinese origin.

日本茶文化虽然源自中国，但经过本土文化的滋润，别具风格。作为日本文化的结晶，日本茶道也是日本文化的最主要代表，集美学、宗教、文学及建筑设计等为一体，重视通过茶事活动来修身养性，达到一种人与自然和谐的精神意境。

Japanese tea culture originates from China, but it has its own style after being nourished by native cultures. Japanese tea ceremony, as the crystal of Japanese culture, is also the most important representative of Japanese culture. It is an assemblage of aesthetics, religion, literature and architectural design. It attaches importance to self-cultivation through tea activities, thus achieving a spiritual conception of harmony between man and nature.

图4.6　圆尔像
Portrait of Enn Ni

图4.7　南浦绍明像
Portrait of Nanpo Syoumyou

● 千利休与待庵
Sen no Rikyū and His Tai-an Tearoom

千利休（1522—1591年），日本战国时代安土桃山时代著名的茶道宗师，日本人称其为"茶圣"。1585年丰臣秀吉在皇宫开设茶会，千利休向正亲町天皇献茶，因此天皇赐予"利休"之居士号。

Sen no Rikyū (1522 – 1591), a famous tea ceremony master in the Azuchi-Momoyama period of the sengoku period, is known as the tea sage in Japan. In 1585, Toyotomi Hideyoshi held a tea party at the imperial palace. Sen no Rikyū offered tea to Emperor Oogimachi, so the latter bestowed him the title of "Rikyū".

待庵是千利休的自作茶室，属于"二叠茶室"，即除去床间，茶室面积为二叠榻榻米。待庵的重要价值在于它是日本茶室的"原型"。

Tai-an Tearoom was made by Sen no Rikyū. It is the "two-tatami tea room", whose size is two tatamis mats except the niche. The important value of Tai-an Tearoom lies in that it is the prototype of the Japanese tea room.

图 4.8 待庵外景
Tai-an Tearoom (outdoor)

图4.9 千利休像
Portrait of Sen no Rikyū

图4.10 待庵内景
Tai-an Tearoom (indoor)

● 和敬清寂
Harmony, Respect, Peace and Tranquility

"和敬清寂"乃茶道之基本精神。"和"指人与大自然之调和;"敬"指由主客之间互相尊敬开始,以至对任何事物都抱有谦敬之心;"清"指心无杂念,令心意纯朴清静,达致"禅"的意境;"寂"更是与大自然融合为一,无始无终之宁静感觉。

"Harmony, Respect, Peace and Tranquility" are the basic spirit of tea ceremony. Harmony refers to the harmony between man and nature. Respect happens between host and guest, and by extension respect for everything. Peace means no distractions, making the mind simple and quiet, and reaching the artistic conception of Zen. Tranquility is a sense of integrating with nature in a peaceful way.

● 利休七则
Seven Rules of Sen no Rikyū

利休七则为茶道宗师千利休的弟子依据其平日教导,所记录下的七条规则:

The Seven Rules of Sen no Rikyū are his daily teachings recorded by his disciples:

茶要泡的合宜入口

炭要好让水滚沸

花的装饰要如在野外般自然

准备好冬暖夏凉的茶室

在预定的时间要提早准备

非下雨天仍要备好雨具

体贴同行客人的心意

Brew tea appropriately for the mouth

Boil water with charcoal

The decorated flowers as natural as in the wild

Prepare a tea room warm in winter and cool in summer

Prepare before scheduled time

Prepare rain gears even in non-rainy days

Be considerate of fellow guests

● "三千家"
"Three Sen Houses"

经过六七百年的漫长岁月,日本茶道发展出众多流派,比较重要的流派现有以千利休为流祖的"三千家",即里千家、表千家和武者小路千家。

Many schools have developed in Japanese tea ceremony over 600 or 700 years. The current important schools include "Three Sen Houses", namely, Urasenke, Omotesenke and keMusha no Koji senke, with Sen no Rikyū as their ancestor.

图 4.11 里千家今日庵
Tea room of Urasenke

图 4.12 表千家不审庵
Tea room of Omotesenke

2. 韩国 Korea

韩国茶文化中最具特色的是茶礼。新罗统一朝鲜半岛（668年）后，茶通过僧侣往来传入朝鲜半岛。高丽时代（918—1392年），茶礼正式成为国家的重要礼仪之一，并传承至今，其基本精神为"和、敬、俭、真"。韩国茶礼种类繁多，有"叶茶法""五行茶礼""成人茶礼""接宾茶礼""佛门茶礼""君子茶礼""闺房茶礼"等。

Tea ceremony is the most distinctive part of Korean tea culture. As Silla unified the Korean Peninsula in 668, tea was introduced to the Korean Peninsula by monks. During the Goryeo period (918 – 1392), tea ceremony officially became one of the important national rituals and has been passed down to this day. Its basic spirit is "harmony, respect, frugality and sincerity". South Korea has many tea ceremonies, including ceremonies of leaf tea, five phases, coming-of-age, guest receiving, Buddhism, gentleman, boudoir, etc.

图4.13 韩国最早试种茶树的智异山华严寺
Jirisan, where the first tea tree was planted in South Korea

漢山州瓢川縣今未詳○按古瓢盧河在今麻田積城之間瓢川或因瓢盧河得名歟有
妖人自言有速富之術衆頗惑之王聞之曰執左道
以惑衆者刑之先王之法也流其人於遠島
冬十二月遣使入朝于唐
遣金大廉入唐帝召對于麟德殿賜宴有差
種茶于地理山
東方初無茶善德王時始有之至是入唐使大廉持
茶種來王使植地理山自此盛焉
以金陽爲固城郡太守
陽字魏昕周元之曾孫也祖宗基襲封溟州王父貞

祖父母忌日除庶人外文武入仕人並給暇
一日兩宵 文宗三十年六月制先亡有後
之妻及同居妻父母忌日依制給暇 明宗
十四年七月制文武入流以上者妻父母忌
日依外祖父母式一日兩宵給暇 恭讓王
三年五月庚子教近設家廟既令六品以上
祭三代自今許行曾祖考妣忌日之祭

重刑奏對儀

王便服出坐內殿南廊承龍都知肅拜訖承
高麗史卷六十四 三十
宣重房院房六尚員次次肅拜閤門去靴笏
著絲鞋侍臣皆同入庭橫行自喝再拜行頭進步
復位各祇候後左邊立刑部奏對員省郎丹
筆員入庭舍人喝再拜出外祇候引宰臣樞
密至門執禮傳引就褥位立定執禮徵喝再
拜執禮承傳云賜座執禮徵喝再拜引上殿
坐東邊褥位茶房參上員從夾戶入進茶內
侍七品員去盖子執禮上殿前楹外面拜勸
茶放後下殿次院房八品以下進宰樞茶執

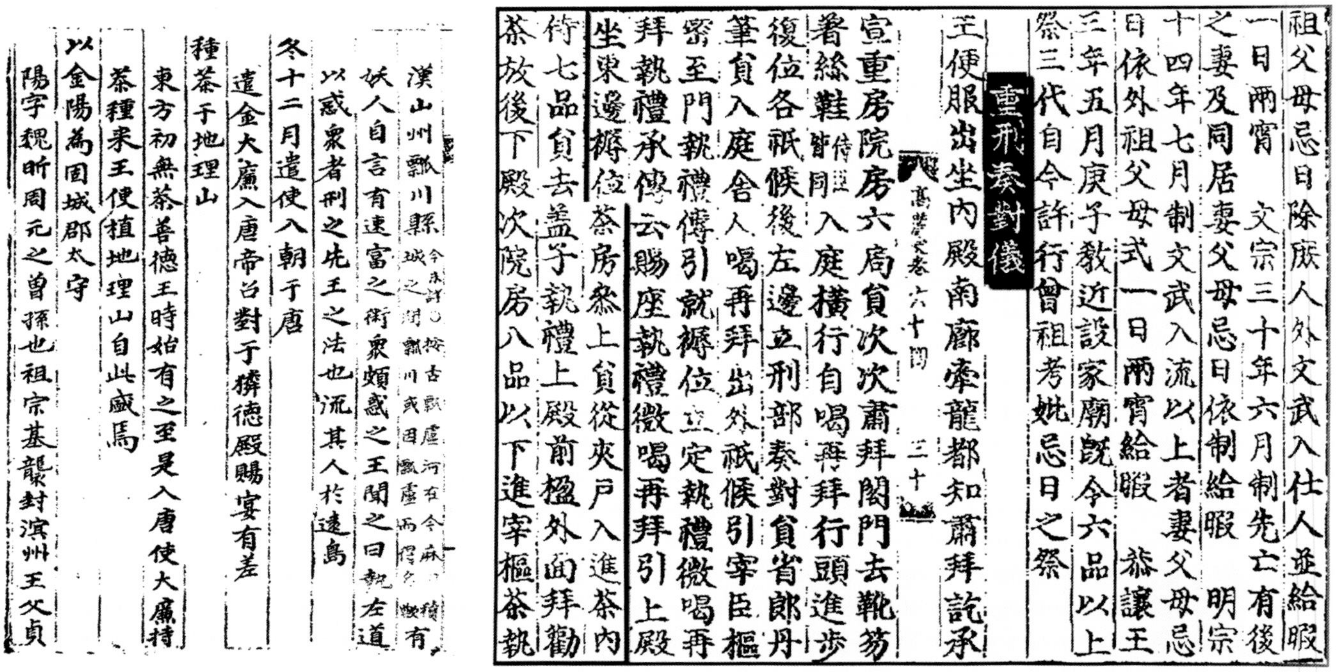

图4.14 韩国古书中记载种茶之事
Records of tea planting in Korean ancient texts

•韩国茶与茶具

Tea and Tea Ware of the Republic of Korea

韩国茶叶大多根据产地和茶叶外形来命名，主要有雀舌茶、竹露茶、紫笋茶、甘露茶等。

Most Korean teas are named after their origin and shape, mainly including bird's tongue tea, bamboo dew tea, purple bamboo shoot tea, and sweet dew tea.

在众多的韩国茶具中，最负盛名的当属高丽青瓷茶具。高丽人崇尚翡色，在各种茶礼中都喜欢使用精美的青瓷茶具。

Korean celadon tea wares are the most famous among all tea sets in the country. Goryeo people admired jade color and liked to use exquisite celadon tea sets in various tea ceremonies.

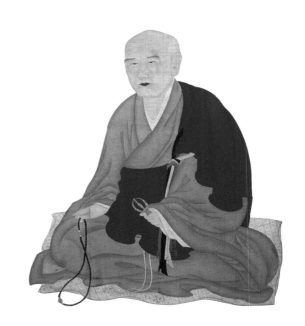

图4.15 韩国茶礼开拓者草衣禅师
Zen Master Caoyi, the pioneer of Korean tea ceremony

图 4.16　韩国茶礼
The Korean tea ritual

3. 印度 India

18世纪之后，印度成为世界主要茶叶生产国和出口国。如今，印度所产的茶叶以阿萨姆和大吉岭为代表。印度人喜欢喝调饮红茶，通常在茶汤中添加香料、砂糖和牛奶等。

After the 18th century, India became the world's leading tea producer and exporter. Today, the representative teas produced in India are Assam and Darjeeling. Indians like drinking mixed black tea, and they usually add spices, sugar and milk to tea soup.

图 4.17　印度饮茶
Tea drinking in India

图4.18 印度茶园
Indian tea garden

4. 斯里兰卡 Sri Lanka

斯里兰卡是目前世界主要产茶国之一。斯里兰卡原以产咖啡闻名，19世纪中叶，茶叶后来居上，取代咖啡成为其支柱产业。

Sri Lanka is one of the world's leading tea producers. The country was once famous for its coffee production, but tea replaced coffee as its mainstay industry in the mid-19th century.

● 斯里兰卡茶园
Sri Lanka Tea Garden

斯里兰卡主要出产红茶，其红茶与印度的大吉岭茶、阿萨姆茶和中国的祁门红茶并称为"世界四大红茶"。斯里兰卡的茶叶按生长的海拔高度分为三类：高地茶、中段茶和低地茶。因海拔高度、气温、湿度的不同，所产茶叶各具特色。

Sri Lanka mainly produces black tea. Its black tea, together with India's Darjeeling Tea, Assam Tea and China's Keemun Black Tea, is called "the world's four major black teas". Tea there can be divided into three categories based on the altitude of growth: High Grown, Middle Grown, Low Grown. Since altitude, temperature and humidity varies, the tea produced has its own characteristics.

图 4.19　斯里兰卡地标与装饰背景
Landmark and its decorative backdrop in Sri Lanka

图 4.20　斯里兰卡采茶人
Sri Lankans plucking the tea buds

5. 土耳其 Turkey

土耳其是世界茶叶生产、消费大国，茶在人们生活中不可或缺，无论是大中城市，或是小城镇，到处都有茶馆。土耳其人饮茶以煮饮为主，讲究调制的功夫。

Turkey is a big tea producer and consumer in the world. Tea is indispensable in people's life, and tea houses can be found everywhere in large and medium cities or small towns. Turks mainly drink boiled tea, and pay attention to the skill of blending.

6. 伊朗 Iran

作为古丝绸之路上的南路要站，伊朗茶文化与中国有着渊源关系。饮茶是伊朗人生活上的一大享受。伊朗人喜欢饮茶，尤其是红茶。每当客人到访之时，更是要准备好精美的茶具和托盘，将煮好的红茶奉上，这才算是待客之道。

Iran was the south stop of the ancient Silk Road. Its tea culture is closely related to China. Drinking tea is a great pleasure in Iranian life. Iranians like to drink tea, especially black tea. When guests come, Iranians will take out exquisite tea utensils and trays to serve boiled black tea, which is the way to host guests.

图4.21　土耳其红茶
Turkey black tea

7. 也门 Yemen

也门人的饮茶风俗很早就由中国传入。与其他阿拉伯国家一样，这里人民多以饮红茶为主。也门人煮茶一般会加入炼乳、豆蔻和丁香，格外香甜。

The Yemeni custom of drinking tea was introduced from China a long time ago. Just like other Arab countries, Yemenis drink mostly black tea. They usually brew tea with condensed milk, cardamom and cloves, making it extraordinarily fragrant and sweet.

图4.22　伊朗茶具
Iranian tea wares

二

欧洲茶文化
European Tea Culture

茶叶输入欧洲后，这一来自东方的星星之火没有寂寂熄灭，而是在适宜的条件下慢慢炽热、旺盛，直至燃遍欧洲。饮茶的盛行为欧洲人民的日常饮食和休闲方式提供了又一种美好的选择。欧洲茶文化，因其明快简洁而别有一种风情。

After tea was introduced into Europe, this "spark" from the East did not die out quietly, but gradually became blazing and vigorous under appropriate conditions, until going all over Europe. The popularity of tea drinking provides a nice alternative for the daily diet and leisure of Europeans. European tea culture has a special style because of its simplicity and conciseness.

1. 英国 Britain

1662 年，葡萄牙公主凯瑟琳远嫁英国。这位嗜茶王后把祖国葡萄牙的饮茶习俗一同带到了英国。由于她的推动，饮茶成了英国宫廷生活的一部分。由于王室贵族的引导，茶成为英国人喜爱的饮料，并逐渐形成了独特的英式下午茶。

In 1662, Portuguese princess Catherine married into Britain. The tea-loving queen brought the Portuguese custom of drinking tea to Britain. Her promotion made tea drinking a part of British court life. As the royal family and aristocrats led the fashion, tea gradually became the favorite drink of the British people, and a unique British afternoon tea was gradually formed.

扫码了解欧洲茶文化

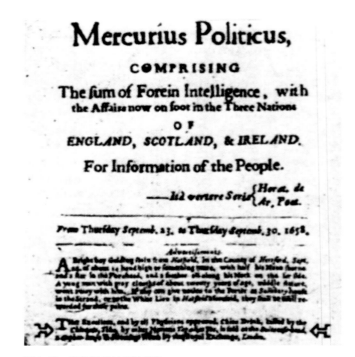

图 4.23　英国最早的茶叶广告
The earliest tea advertisement in Britain

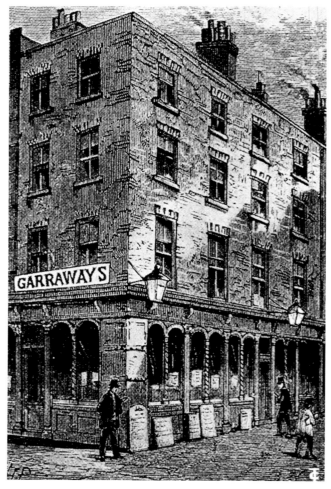

图 4.24　英国最早出售茶叶的加仑威尔咖啡店
The first British coffee house that sold tea

- 英式下午茶

  The British Afternoon Tea

19世纪初期，茶在英国日渐普遍，著名的英国下午茶就出现在维多利亚时代（1837—1901年）。斐德福公爵夫人安娜·玛丽亚常于下午四点左右邀请朋友喝茶聊天，之后这一活动逐渐流行于各阶层。如今，下午茶已成为英国人的一种生活方式和休闲文化。

It was in the early 19 th century, when tea became increasingly popular in the Britain, that the British afternoon tea first appeared. It was also called "the Victorian era (1837 – 1901)" . Legend says that Anna Maria Russell, Duchess of Bedford, often invited her friends to a tea party around 4 p.m., and the party thus started to gain popularity among all social classes. Having afternoon tea, nowadays, is a way of life and culture for the British.

- 英国茶具

  Tea Sets in the Britain

饮茶的流行引发人们对茶具的需求，当来自东方的精美瓷器一进入欧洲，便成为全欧洲上层社会争相购买的奢侈品。后来，英国的手工艺人开始仿制中国茶具，逐渐形成自己的风格。

The popularity of tea sparked public demand for tea sets. The moment the exquisite porcelain from China landed in Europe, it soon became luxury goods that the aristocrats in Europe rushed to purchase. The unique charms of the British tea sets, as an imitation of the Chinese ones, began to take shape by that time.

- 茶叶与英国东印度公司

  Tea and British East India Company

英国东印度公司成立于1600年，至19世纪30年代，它垄断了西方与中国茶叶的贸易往来。在茶叶贸易中获得的巨大利润，是英国重要的收入来源。

The British East India Company, founded in 1600, had monopolized the tea trade between the West and China till the 1830 s. The company made a huge profit in the tea trade, which was an important source of income for Britain.

1840—1929 年英国年人均茶叶消费量 5 年平均

Annual per capita tea consumption in Britain between 1840 and 1929

年份	数量（磅）
1840—1844	1.39
1845—1849	1.70
1850—1854	2.04
1855—1859	2.45
1860—1864	2.79
1865—1869	3.83
1870—1874	4.01
1875—1879	4.56
1880—1884	4.71
1885—1889	5.00
1890—1894	5.37
1895—1899	5.79
1900—1904	6.06
1905—1909	6.22
1910—1914	6.58
1915—1919	7.18
1920—1924	8.64
1925—1929	9.23

数据来源：[美]威廉·乌克斯：《茶叶全书》。

Source: [USA] William H. Ukers, *All About Tea*.

图 4.25　广州珠江沿岸十三行

The Thirteen Hongs in Guangzhou

• 茶叶与鸦片战争
Tea and "Opium War"

19世纪，在中英贸易中，英国对茶叶的需求量日益增加。为了弥补同中国的贸易逆差，英国利用印度殖民地全力发展对华鸦片贸易，由此引发了中英"鸦片战争"。

In the 19th century, Britain had an increasing demand for tea in the Sino-British trade. In order to make up for the trade deficit with China, Britain fully develop its opium trade with China in its Indian colony, thus triggering the "Opium War" between China and Britain.

2. 俄罗斯 Russia

茶从16世纪开始传入俄国及东欧诸国，17世纪后期，饮茶风习已传播到各个阶层。19世纪时，俄国出现了许多记载茶俗、茶礼、茶会的文学作品。如俄国著名诗人普希金就有俄国乡间茶会的记述。现今，俄罗斯的茶室遍布都市、城镇及乡村。俄罗斯人爱饮红茶，并习惯加糖、柠檬片或牛奶，还伴以蛋糕、烤饼、甜面包等茶点。茶炊即茶汤壶，是俄罗斯人饮茶必备的茶具，两层，壁四围灌水，在中间加热。通常为铜质，有球形、桶形、花瓶形、罐形等。

Beginning in the 16th century, tea was introduced into Russia and Eastern European countries. By the late 17th century, tea drinking customs had spread to all classes. In the 19th century, many Russian literary works gave an account of tea customs, ceremonies and parties. For example, Pushkin, a famous Russian poet, described Russian country tea party. Today, Russian tea houses can be found in cities, towns and villages. Russians love to drink black tea and tend to add sugar, lemon slices or milk in it, accompanied by refreshments such as cakes, scones and sweet bread. The two-layer samovar, namely the tea pot, is a necessity for Russians to drink tea. Water is filled and heated via a pipe that runs through the middle of the vat. It is usually copper-made and variously shaped, including sphere, barrel, vase, pot and so on.

图 4.26 鲍里斯·库斯托季耶夫《商人妻子的茶》
Merchant's Wife at Tea by Boris Kustodiev

叶普盖尼·奥涅金(节选)
Eugene Onegin (Excerpts)

黄昏来临,烧晚茶的茶炊,

在桌上闪光,咝咝作响,

热着中国茶壶里的茶水,

轻轻的水汽在它下面飘荡。

奥丽加亲手给大家倒茶,

香气馥郁的茶水像一股

浓黑的水流斟入了茶碗。

小厮还双手送上了凝乳;

达吉雅娜呆呆站在窗前,

对着冰冷的玻璃窗呼吸,

我的宝贝,她默默地想着心事,

轻轻地划动着娇嫩的手指,

在蒙着雾气的玻璃窗上面,

写上心爱的奥和叶两个字。

As dusk approaches, the samovar for evening tea

flashes and hisses on the table,

heating the tea in the Chinese teapot,

and gentle vapor floats beneath it.

Olga poured tea for everyone by herself.

The fragrant tea water poured into the tea bowl

like a stream of thick black water.

The manservant offered curd;

Tatyana stood in front of the window,

breathing into the cold glass window.

My darling, she brooded in silence,

gently moved her delicate fingers,

and wrote the words "Ao" and "Ye"

on the misty glass window.

——普希金

Pushkin

3. 法国 France

茶叶于17世纪传入法国。18世纪，饮茶开始流行于法国社会各阶层。最早进入法国的茶叶是绿茶，随后有乌龙茶、红茶、花茶及沱茶。如今，法国人所饮的茶叶主要是红茶，饮茶方式主要有"清饮"和"调饮"。"清饮"和中国目前的饮茶方式相似，"调饮"则加方糖或新鲜薄荷叶，使茶味甘甜。

Tea was introduced to France in the 17th century. In the following century, tea became popular among all orders of France. Green tea was first introduced into France, followed by Oolong tea, black tea, scented tea and tuocha. Now the French mainly drink black tea, which is made either by adding sugar or fresh mint leaves to make the tea sweet, or adding nothing else. The latter is similar to the Chinese way of drinking tea.

图4.27 巴黎街头茶壶形的沙龙标志
The teapot shaped salon logo on the street of Paris

图4.28 法国贵族饮茶场景
French nobles drinking tea

4. 德国 Germany

茶叶大约于17世纪中期传入德国。1757年，普鲁士国王腓特烈二世在波茨坦市北郊的无忧宫园林内，特地修建了一座具有中国风格的中国茶亭。如今，喝茶已经越来越受到德国人的青睐，其中红茶消费量最高，其次是绿茶。

Tea was introduced into Germany around the mid-17th century. In 1757, Frederick II, King of Prussia, built a Chinese style tea house in the garden of Sanssouci Palace in the northern suburb of Potsdam. Today, tea drinking has become increasingly popular in Germany. Black tea is consumed the most, followed by green tea.

图4.29　德国茶树
German tea tree

图4.30　波茨坦无忧宫里的中国茶亭
Chinese style tea house in the garden of Sanssouci Palace of Potsdam

5. 瑞典 Sweden

瑞典与中国茶叶的最早联系可以追溯到18世纪，其历史见证就是"哥德堡号"远洋商船。1763年瑞典植物学家林奈在瑞典的乌普萨拉率先试种茶树，并获得成功，但因为气候原因并未普及。由于价格昂贵，只有上层社会的贵族才会大量饮茶。直到袋泡茶的出现，茶叶才开始真正在瑞典普及。如今，瑞典人会饮用各种茶，但红茶始终是最流行的。

The connection between Sweden and Chinese tea can be traced back to the 18th century, which was witnessed by the ocean-going merchant ship "Gotheborg". In 1763, the Swedish botanist Linnaeus first planted tea trees in Uppsala, Sweden, and it was a success. But the climate made it impossible to plant tea widely. Because of the high price, only the upper-class nobles drank a lot of tea. Tea was not really popularized in Sweden until the emergence of tea bags. Nowadays, Swedes drink a variety of teas, but black tea remains the most popular.

6. 芬兰 Finland

芬兰人日常饮料以咖啡为主，但是爱饮茶者也不乏其人，尤其是18—19世纪时东部地区茶叶消费量曾一度超过咖啡。现在芬兰茶叶市场品种颇多，以红茶为大宗，绿茶仅占10%左右，另外还有各种果味茶。芬兰人虽然饮茶不多，但是很有特色。至今还有不少芬兰人使用一种叫"沙莫瓦尔"的俄罗斯式铜茶壶煮茶，加上专门的茶具和饮茶礼仪，文化气息浓郁。

In Finland, though coffee is the main daily drink, a large number of people love tea, especially in the eastern part of the country, where tea consumption exceeded coffee during the 18th and 19th centuries. At present, there are many kinds of tea in the Finnish market, most of which are black tea. Green tea accounts for only about 10%, and there are also various fruit-flavored teas. The Finns do not drink much tea, but their tea drinking is very distinctive. Many Finns still make tea in a Russian style copper teapot called "samovar". With special tea sets and tea drinking etiquette, tea drinking in Finland has a strong sense of culture.

图4.31　林奈
Linnaeus

图 4.32 瑞典乌普萨拉植物园
Uppsala Botanic Garden in Sweden

三

非洲茶文化
African Tea Culture

非洲，一片热情、遥远、古老、神秘的土地。如果要为非洲画一幅速写，炽热的阳光、无垠的沙漠、茂密的雨林、纵横的河流，或许是不可缺少的线条。而葱茏的茶园和清甜的茶饮，也值得仔细勾勒，它们滋养了非洲人民，舒展了他们的身心。

Africa is a warm, remote, ancient and mysterious land. A sketch of Africa can't be finished without the hot sunshine, boundless desert, dense rainforest and crisscross rivers. Verdant tea gardens and sweet teas also need being depicted. They nourish the African people and refresh them physically and spiritually.

扫码了解非洲茶文化

图4.33 埃及街头饮茶
Drinking tea on the streets of Egypt

1. 埃及 Egypt

埃及是非洲茶叶进口和消费的主要国家。埃及人好饮红茶，采用煮饮的方式，喜欢在茶中加糖，调成香甜甘醇的糖茶。

Egypt is a major tea importer and consumer. Egyptians like to drink boiled black tea. Sugar is often added into the tea to make it sweet and mellow.

图4.34 埃及茶
Egyptian tea

2. 摩洛哥 Morocco

摩洛哥是北非地区仅有的绿茶消费国，茶对于摩洛哥人的重要性仅次于吃饭。摩洛哥是世界上进口绿茶最多的国家，大部分茶叶从中国进口，仅在其北部的丹尼尔地区有少量的茶叶生产。摩洛哥人喜欢喝浓茶，不仅茶叶量大，薄荷和糖也加得多。当地人认为，只有本地出产的糖，才能泡出最好的茶。

Morocco is the only green tea consumer in North Africa, and the importance of tea is only second to meals in Morocco. Morocco is the world's largest importer of green tea, and most of its tea is imported from China. Only a small amount of tea is produced in the Tangier region in the north of Morocco. Moroccans like to drink strong tea, with a large amount of tea, mint and sugar. Locals believe that the best tea can only be made with locally grown sugar.

摩洛哥的茶具自成风格，茶具制作为当地一绝。摩洛哥的铜器制造业发达，其茶壶一般用铜锤打而成，然后镀银，壶身还錾刻伊斯兰风格的纹饰。传统设计的镀银铜茶壶和托盘，是摩洛哥人饮用传统薄荷茶时常常使用的茶具。在节日或亲朋好友聚会时，这样的茶具必不可少。

Moroccan tea sets have their own style, and production of tea sets is a local wonder. Morocco has a well-developed copper ware manufacturing industry. Its teapots are usually hammered out of copper and plated with silver, and the body is chiseled with Islamic patterns. Silver-plated copper teapots and trays of traditional design are often used by Moroccans when drinking traditional mint tea. Such tea sets are indispensable in festivals or gatherings with friends and relatives.

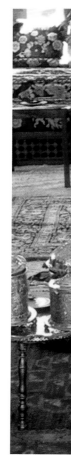

图 4.35　薄荷茶
Mint tea

图 4.36　摩洛哥人饮茶
Moroccans drinking tea

3. 肯尼亚 Kenya

位于东非的肯尼亚是世界主要茶叶生产国，植茶区域遍布全国，海拔均在1000～2700米之间，年平均气温为21℃左右。光照充足，雨量丰富，略带酸性的火山灰土壤极其肥沃，非常适宜茶树的生长。肯尼亚茶文化极其丰富。肯尼亚人受英国人的影响，有喝下午茶的习惯。民众主要饮用红碎茶，一般会在红茶中加糖调饮。

Kenya in East Africa is a major tea producer in the world. Tea planting areas are distributed all over Kenya, with an altitude of 1000 – 2700 meters and an annual average temperature of about 21 ℃. The sufficient light, abundant rainfall and extremely fertile slightly acidic volcanic soil are very suitable for the growth of tea trees. Kenya is extremely rich in tea culture. Influenced by the British, Kenyans are in the habit of drinking afternoon tea. They mainly consume broken black tea, which is usually mixed with sugar.

图4.37 非洲人喜爱饮用的柠檬绿茶
Lemon green tea loved by Africans

图4.38 肯尼亚红茶包
Kenyan black tea bags

图4.39 肯尼亚采茶人
Tea pickers in Kenya

四

美洲茶文化
American Tea Culture

在大洋彼岸的美洲，茶叶曾经引发了一场争取独立与自由的战争。小小的叶子掀起的波澜，一重重推进，最终影响了国家的命运。硝烟早已散尽，美洲人民对茶叶的热情不减，如今，他们依然以自己的方式理解茶、利用茶、亲近茶，延续美洲的茶文化。

Tea once sparked a war for independence and freedom in America. The waves raised by small leaves pushed forward one after another, and finally affected the fate of the country. The smoke of gunpowder has long gone, and the American people's passion for tea continues unabated. Today, they still understand tea, use tea and get close to tea in their own way, thus continuing the American tea culture.

1. 美国 The United States

北美与茶的渊源可追溯至17世纪。远在1690年，波士顿就有定点出售中国茶叶的市场，1773年，因拒绝上缴增加的茶税而引发的"波士顿倾茶事件"成为美国独立战争的导火索。美国独立之后，于1784年派遣第一艘"中国皇后号"商船，开始了中美贸易，其中茶叶是重要的货物。18世纪60年代，美国进口的茶叶已经达到120万磅。

The origin of tea in North America can be traced back to the 17th century. There was a designated market for selling Chinese tea in Boston as early as 1690. In 1773, the "Boston Tea Party" provoked by the refusal to pay the increased tax on tea became the trigger for the American Revolution. After the United States won its independence, the first merchant ship "Empress of China" was dispatched to start Sino-American trade in 1784, in which tea was an important commodity. By the 1760s, the United States had imported 1.2 million pounds of tea.

- 波士顿倾茶事件
  Boston Tea Party

波士顿倾茶事件，又被称作波士顿茶会事件。1773年，英国国会通过《茶税法案》，这个税法引起了英国殖民地人民的强烈反对。1773年12月16日，东印度公司满载中国茶的船到达波士顿港口。殖民地人民拒付茶税，兴起抗税运动，数千民众向船只涌去，其中一些伪装成了美洲印第安人。抗议者们把342箱茶叶抛入海中。此举成为1775年美国独立战争的导火线。

The Destruction of the Tea in Boston is also known as the Boston Tea Party. In 1773, the British Parliament passed the *Tea Act*, which provoked fierce opposition from the people of British colonies. On December 16, 1773, ships of East India Company fully loaded with Chinese tea arrived at Boston Harbor. The American colonists protested against taxes on tea. Thousands of people swarmed into the British ships, some of them disguised as American Indians. They dumped contents of 342 chests of tea into the harbor. The incident led to the outbreak of the American Revolution in 1775.

- "中国皇后号"
  Empress of China

"中国皇后号"是一艘三桅帆船，1784年2月22日自纽约启航，5个月后抵达广州，成为美国立国后第一艘抵达中国的船只。装满茶叶后，"中国皇后号"于次年5月11日返回纽约港。这次航行是中美历史上的首次通商，开辟了两国的贸易之门，亦鼓舞了美国的商人和水手们。

The Empress of China was a barque that set sail from New York on February 22, 1784 and arrived in Guangzhou five months later, becoming the first ship to arrive in China after the founding of the United States. After being filled with tea, the Empress of China returned to New York Harbor on May 11 in the following year. This voyage witnessed the first commercial intercourse between China and the United States in history, opening the door to trade between the two countries and inspiring American merchants and sailors.

扫码了解波士顿倾茶事件

图4.40 波士顿倾茶事件的商船
The ships in the Boston Tea Party

图4.41 波士顿码头
A wharf in Boston

- 威廉·乌克斯所著《茶叶全书》
  All About Tea by William H. Ukers

该书为作者历时30余年搜集资料、访问众多产茶国家和地区，写作10余年而成，于1935年出版。全书共60余万字，对茶叶的各个方面及主要产茶国家和地区都有丰富详尽的记述，堪称20世纪的茶学巨著。

The book, published in 1935, was written for more than 10 years after the author spent more than 30 years collecting materials and visiting many tea-producing countries and regions. With a total of more than 600,000 words, the book described all aspects of tea and the main tea-producing countries and regions in detail, and it can be called a masterpiece of tea science in the 20th century.

- 饮茶习俗
  Tea Drinking Custom

美国人最早饮用的是绿茶。20世纪初，随着冰茶的出现，红茶在北美的销售量超过了绿茶。1920年，红茶和绿茶各占40%，乌龙茶占20%。后来，红茶成为美国人的首选。

Americans first drank green tea. With the emergence of iced tea in the early 20th century, the sales volume of black tea surpassed that of green tea in North America. In 1920, black tea and green tea accounted for 40% respectively, and oolong tea 20%. Later, black tea became the first choice of Americans.

美国市场上的东方茶，诸如绿茶、乌龙茶等茶类品种很多，但饮的多是罐装的冷饮茶，尤以柠檬绿茶和柠檬红茶为多。美国人与中国人饮茶方法不同，不喜欢热饮或温饮，更喜欢冷饮。

There are a variety of Oriental teas in the American market, including green tea, oolong tea, etc., but Americans drink mostly canned cold tea, especially lemon green tea and lemon black tea. Unlike the Chinese way of drinking tea, Americans prefer cold drinks rather than hot or warm ones.

图 4.42　美国家庭饮茶
The family tea party in the United States

图4.43 马黛茶树
Yerba mate

2. 阿根廷 Argentina

阿根廷人有饮用"南美仙草"——马黛茶的习惯。所谓马黛茶,其实是一种"非茶之茶"。马黛树一般株高3~6米,野生的可达20米,树叶翠绿,呈椭圆形,枝叶间开雪白小花。阿根廷人对这种叶子的处理方法和中国的茶叶相似,所以中国把这种叶子称为马黛茶。

Argentinean has the habit of drinking yerba mate tea, known as the "South American herb". Yerba mate tea is actually not a tea. Yerba mate is generally 3–6 meters tall, and the wild tree can be 20 meters tall. The leaves are green and oval, with small snow-white flowers between the branches and leaves. Argentinean treats the leaves in a similar way to tea in China, so the leaves are called yerba mate tea in China.

图4.44 马黛茶具
Yerba mate teaware

- 马黛茶文化
  Yerba Mate Tea Culture

阿根廷是全球最大的马黛茶生产国，马黛茶不仅是当地人民生活中不可缺少的饮料，而且大量出口北美、西欧和日本等国。

Argentina is the world's largest producer of yerba mate tea, which is an indispensable drink in local people's life, and is also exported in large quantities to North America, Western Europe, Japan etc.

图 4.45　阿根廷马黛茶具
Yerba mate teaware

马黛茶具有极高的营养价值，其中含有蛋白质、碳水化合物、淀粉、维生素C、维生素B1、维生素B6、镁、铁、钾、钙、磷等多种微量元素。马黛茶也具有药用价值，因此被称为"健康之饮""快乐之茶"，成为阿根廷的一种标志文化。

Containing protein, carbohydrate, starch, vitamin C, B1, B6, magnesium, iron, potassium, calcium, phosphorus and other trace elements, yerba mate tea is of high nutritional values. It also has medicinal values, so it is hailed as "healthy drink" and "happy tea", and it has become an icon of Argentine culture.

图 4.46　马黛茶包
Yerba mate teabags

图 4.47　阿根廷人饮用马黛茶
Yerba mate tea for Argentinians

五

大洋洲茶文化
Oceania Tea Culture

大洋洲人民的饮茶习俗由来已久，茶饮在人们的日常生活中分量也重。在饮茶方式上，大洋洲人民形成了自己的偏好。无论饮茶方式和茶味浓淡与其他地方有何异同，在这里，一天之中总有一段或更多的时光留给茶。

The people of Oceania have a long history of tea drinking. Tea is a weighty item in their daily life. They have their own preferences in the way that tea is made. No matter how tea is drank and tastes, there is always time for tea in a single day.

1. 澳大利亚 Australia

澳大利亚不仅产茶，也是茶叶消费大国，所产茶叶远远不能满足国内需求，主要依靠进口。受英国影响，澳大利亚人喜欢红碎茶，红碎茶占消费总量的85%。红茶的饮用为调饮式，常在茶中加入糖、牛奶或柠檬。

Australia is not only a tea producer, but also a large tea consumer. The tea yielded in Australia is far from meeting domestic demand, so the country mainly depends on imports for tea. Affected by the Britain, Australians like broken black tea, which accounts for 85% of the total consumption. They tend to drink tea with sugar, milk or lemon in it.

图4.48　澳大利亚茶园
Australian tea garden

图4.49　澳大利亚茶叶包装
Australian tea packaging

图 4.50 新西兰茶园
Tea plantations in New Zealand

2. 新西兰 New Zealand

新西兰的饮茶风俗与澳大利亚相似，习惯饮红茶，喜爱加糖加奶，甚至加入甜酒、柠檬饮用。新西兰原本不产茶，茶叶消费完全依赖进口。20世纪初，中国人在新西兰开辟了茶园，才开始发展茶叶生产。如今，新西兰已有茶园万余亩，其生产规模还在持续扩大中。

The tea drinking custom in New Zealand is similar to that in Australia. New Zealanders have the habit of drinking black tea with sugar and milk, and even sweet wine and lemon. Initially tea was not grown in New Zealand, so the country had to import all the tea. In the early 20th century, Chinese people opened tea gardens in New Zealand and tea production started. Today, New Zealand has more than 10000 *mu* of tea, and its production scale is still expanding.

图 4.51 新西兰茶叶包装
The tea package from New Zealand

图 4.52 新西兰茶室
The tea house in New Zealand

第三章 世界茶业

Tea Industry in the World

在茶为中国举国之饮的今天，世界范围之内，茶亦被列为世界三大无酒精饮料之首，是全世界人们普遍欢迎的一种天然、营养、保健的绿色饮料。从全球茶区的地理分布看，种茶国家遍及世界五大洲，茶产业早已成为全球农业的重要组成部分。

Today, tea is the national drink of China, and is also listed as the first of top three non-alcoholic beverages in the world. It is a natural, nutritious and healthy green drink widely welcomed worldwide. As shown in the geographical distribution of tea regions all over the world, tea growing countries are distributed throughout the five continents. Tea industry has become an important part of global agriculture.

世界茶业大事记
A Chronicle of Events of the Global Tea Business

- 公元前 4000—前 3500 年

浙江余姚人工种植茶树。

4000 BCE – 3500 BCE

Tea trees were planted in Yuyao, Zhejiang.

- 公元前 59 年

王褒订立《僮约》。

59 BCE

Wang Bao made the *Master-Slave Contract*.

- 约 250—305 年

晋代左思写作《娇女诗》。

Ca. 250 – 305

Zuo Si in the Jin Dynasty wrote the *Poetry of Beloved Daughters*.

- 348—355 年

《华阳国志》载三千多年前四川一带已种茶。

348 – 355

According to the *Record of Huayang State*, tea was planted in Sichuan more than 3000 years ago.

- 456—536 年

陶弘景《杂录》云："苦茶轻身换骨，昔丹丘子、黄山君服之。"

456 – 536

It was recorded in the *Miscellaneous Records* by Tao Hongjing, "Bitter tea makes people feel comfortable and energetic. Danqiuzi and Huangshanjun drank it in the early days."

- 483—493 年

齐武帝萧赜下诏"以茶为祭"。

483 – 493

Xiao Ze, the Emperor of Qi, issued an imperial edict to "sacrifice with tea".

- 770 年

唐政府在宜兴、长兴顾渚山开始设立贡茶院。

770

The Tang government set up tribute tea workshops in Guzhu Mountain between Yixing and Changxing.

- 780 年

陆羽编撰了世界第一部茶叶专著——《茶经》。

唐政府开始征收茶税。

780

Lu Yu compiled *The Classic of Tea*, the world's first monograph on tea.

The Tang government began to levy a tax on tea.

- 805 年

日本最澄和尚把第一批茶籽从中国带回日本。

805

Saichō, a Japanese Buddhist monk, brought the first batch of tea seeds from China to Japan.

- 828 年

新罗使节金大廉，将从中国带回的茶籽种于智异山下，揭开了朝鲜半岛的种茶史。

828

An envoy of Silla planted the tea seeds brought back from China at the foot of Jirisan, starting the history of tea planting on the Korean Peninsula.

● 835 年

唐政府开始建立"榷茶"制度。

835

The Tang government began to levy a tax and acquire a monopoly on tea.

● 880 年

《中国印度见闻录》成书，书中有阿拉伯人对中国茶的最早记载。

880

A Voyage to China and India was compiled, which contains the earliest record of Chinese tea by Arabs.

● 951 年

在日本茶被用来抵御鼠疫。

951

Tea was used to combat the plague in Japan.

● 960 年

宋太祖下诏设茶库。

960

Emperor Taizu of the Song Dynasty issued an imperial edict to set up a tea storehouse.

● 977 年

宋太宗在建安设官焙，专造北苑贡茶，龙凤团茶大发展。

977

Emperor Taizong of the Song Dynasty set up the official workshops in Jian'an, making Beiyuan tribute tea. Dragon-phoenix cake tea greatly developed.

● 1074 年

北宋设立茶马司。

1074

The Tea-Horse Department was established in the Northern Song Dynasty.

● 1085 年

高丽国僧统义天首次入宋，回国后把中国的茶禅一味精神带到高丽，并引进宋皇室专用的龙凤团茶。

1085

Chief Monk of Goryeo Kingdom, Yitian, came to Song China for the first time. When he returned home, he brought with him the Chinese spirit of tea and Zen affinity, and dragon-phoenix cake tea for the imperial use.

● 1107 年

宋徽宗《大观茶论》问世。

1107

The Treatise on Tea from the Daguan Reign Period by Song Huizong came out.

● 1187 年

日本高僧荣西再来中国学佛，回国时带走茶种，种于京都脊振山下。

1187

Eisai, a Japanese eminent monk, revisited China to study Buddhism. When he went back to Japan, he took some tea seeds and planted them at the foot of a mountain in Kyoto.

● 1211 年

日本第一本茶叶专著——《吃茶养生记》问世。

1211

The first Japanese treatise on tea, *An Account of Drinking Tea for Nourishing Life* was finished.

● 1391 年

明太祖下诏废除团茶，改贡散茶。

1391

Emperor Taizu of the Ming Dynasty issued an edict to ban dragon-phoenix cake tea as tribute. Instead, loose tea was collected.

- 1445 年

朱权撰写《茶谱》，较系统地记述了蒸青散茶的烹饮之法。

1445

Zhu Quan wrote the *Tea Manual*, which systematically describes the boiling and drinking methods of steamed loose tea.

- 1607 年

荷兰东印度公司的船从爪哇运中国茶到欧洲。

1607

The ship of Dutch East India Company transported Chinese tea from Java to Europe.

- 1618 年

中国遣使入俄，向沙皇赠茶。

1618

Qing China sent envoys to Russia to present tea to the tsar.

- 1650 年

法、俄等国家的商人相继来中国订购瓷茶具。

1650

Businessmen from France, Russia and other countries successively came to China to order porcelain tea sets.

- 1655 年

荷兰人到广州购茶，中方宴请时用牛奶掺入茶中作饮品。荷兰人尼荷之后在著作中指出："西方奶茶之起，即源于此。"

1655

Dutch people came to Guangzhou to purchase tea. The Chinese hosts added milk into tea as a drink at the banquet. A Dutchman later pointed out in his works, "Western milk tea originated from this."

- 1657 年

中国茶叶在法国市场销售。英国伦敦加仑威尔咖啡店开始售卖中国茶。

1657

Chinese tea was sold in France. In London, England, Garraway's Coffee House began to sell Chinese tea.

- 1658 年

英国伦敦《政治公报》刊登西方最早宣传中国茶的广告。

1658

The first advertisement promoting Chinese tea in the West was published on the *Political Gazetee* in London, England.

- 1663 年

英王查理二世的皇后葡萄牙凯瑟琳公主生日时，英国诗人沃勒为其撰写西方第一首茶诗，以示庆贺。

1663

English poet E. Waller wrote the first tea poem in the West to celebrate the birthday of Princess Catherine of Portugal, Queen of King Charles II of England.

- 1673 年

中国的瓷茶具销往当时的巴达维亚、马六甲、柔佛等地。

1673

Chinese porcelain tea wares were sold to Batavia, Malacca, Johor and other places.

- 1684 年

德国人安德烈亚斯·席勒将茶树从日本传到爪哇岛，种植在雅加达植物园，供游人观赏。

1684

A German with the name of Andreas Schiller introduced tea trees from Japan to Java Island and planted them in Jakarta Botanical Garden as a treat to sightseers.

● 1688 年

北美的英国人从澳门购茶 120 吨运往纽约。

1688

The British in North America bought 120 tons of tea from Macao and shipped it to New York.

● 1689 年

福建厦门出口箱茶 150 担到英国，为中国内地与英国茶叶直接贸易之先声。

中俄签订《尼布楚条约》，中国茶源源输入俄国。

1689

Totally 150 *dan* of tea was exported from Xiamen, Fujian to Britain, which was the first direct trade between inland China and Britain.

Qing China and Russia signed the *Treaty of Nerchinsk*, and Chinese tea was continuously imported into Russia.

● 1690 年

中国茶叶输入美国波士顿，获得在波士顿售茶的特许执照。

1690

Chinese tea was imported into the American city Boston and a special license to sell tea there was awarded.

● 1702 年

英商在浙江舟山设贸易站，采购珠茶。

1702

British merchants set up a trading post in Zhoushan, Zhejiang, to purchase gunpowder tea.

● 1732 年

瑞典开始与中国通商，来广州用粗绒等换取茶叶。

1732

Sweden began its trade with China, exchanging rough wool and other products for tea in Guangzhou.

● 1752 年

荷兰东印度公司"葛尔德马尔森"号远洋轮从广州满载茶叶和瓷器回航，途中在中国海域触礁沉没。

1752

Fully loaded with tea and porcelain, the seagoing vessel "Geerdemaersen" of the Dutch East India Company returned from Guangzhou, hit rocks and sank in Chinese waters on the way.

● 1760 年

英国东印度公司从广州经澳门等港口运出的茶，年价值超过白银 80 万两。

1760

The tea shipped by the British East India Company via Macao and other ports from Guangzhou was valued at more than 800,000 tael of silver yearly.

● 1768 年

茶为中国输出缅甸的主要货物之一。

1768

Tea was one of the main goods exported from China to Myanmar.

● 1773 年

英当局向美国殖民地强征茶税，引发"波士顿倾茶事件"。

1773

The British authority imposed tax on tea in the American colonies, triggering the Boston Tea Party.

● 1780 年

英国东印度公司船长从广州购得茶籽，运至加尔各答试种。

1780

One captain of the British East India Company bought tea seeds from Guangzhou and shipped them to Calcutta for trial planting.

• 1784 年

美国商船"中国皇后号"首次抵达广州黄埔港，大量购买中国茶，为美国从中国运茶的最早记载。

1784

The American merchant ship "Empress of China" arrived at Huangpu port in Guangzhou for the first time. It purchased large quantities of Chinese tea, which is the earliest record of the United States transporting tea from China.

• 1792 年

恰克图复市，中国茶叶陆路出口再次畅通。

1792

The reopening of trade in Kyakhta revived the overland export of Chinese tea.

• 1802 年

锡兰试种茶树失败。

1802

Trial planting of tea trees failed in Ceylon.

• 1810 年

福建人柯朝携茶籽至台湾，授种茶之法。

1810

Ke Chao, a Fujian native, brought tea seeds to Taiwan and taught people there how to cultivate tea.

• 1823 年

罗伯特·布鲁斯在印度阿萨姆发现野生茶树。

1823

Robert Bruce found wild tea trees in Assam, India.

• 1826 年

爪哇试种西博德博士由日携返之野生茶树。

1826

Wild tea trees that Dr. Siebold brought back from Japan were tentatively planted in Java.

• 1833 年

雅各布森最后一次(第六次)由中国返爪哇，携茶籽700万粒、茶工15人及多种制茶工具。

1833

Jacobson returned to Java from China for the last time (the sixth time), with 7 million tea seeds, 15 tea workers and various tea processing tools.

• 1834 年

印度总督威廉下令组织茶叶委员会，研究印度茶栽植方案。印度茶叶委员会派秘书乔治赴中国罗致茶工，收购茶籽，考察华茶制造方法。

1834

William, Viceroy of India, ordered that a tea board should be established and work on a tea planting program in India. The board sent Secretary George to China to recruit tea workers, purchase tea seeds and investigate the processing methods of Chinese tea.

• 1839 年

伦敦成立茶叶拍卖市场，首开世界茶叶自由拍卖交易。

1839

A tea auction market was set up in London, initiating the free tea auction in the world.

• 1844 年

美国宣布免除中国茶叶进口税。

1844

The United States announced the exemption of import duties on Chinese tea.

• 1847 年

俄外高加索开始种茶。

1847

Russian Transcaucasia began to grow tea.

- 1851 年

上海出口茶叶超过广州，成为中国茶叶主要输出港口。

1851

Shanghai surpassed Guangzhou in tea exports and became a major export port of tea in China.

- 1859 年

继广州、上海之后，福州的茶叶出口总量占全国第一位。

1859

Fuzhou became the largest tea exporter in China, following Guangzhou and Shanghai.

- 1861 年

俄商在湖北汉口成立第一家砖茶加工厂。

1861

Russian businessmen set up the first brick tea processing plant in Hankou, Hubei.

- 1863 年

俄商在湖北羊楼洞等地开设三家砖茶厂，将湖北收购的茶叶运到俄国各消费城市销售。

1863

Russian businessmen set up three brick tea plants in Yangloudong and other places in Hubei. They transported the tea purchased in Hubei to Russia for sale.

- 1864—1869 年

1864年，英国人杜德在台湾考察时发现台湾北部淡水河流域适宜发展茶叶，遂于1866年设立宝顺洋行，推广种茶，收购茶叶。1868年杜德在今台北万华设精制厂，从此台茶不必运往厦门或福州精制。1869年杜德将台茶直接销往美国。

1864 – 1869

In 1864, the British man John Dodd found that Tamsui River in northern Taiwan was suitable for tea industry while investigating in Taiwan. So he founded Dodd & Company in 1866 to promote tea planting and purchase tea as well. In 1868, Dodd set up a refining plant in today's Taipei Wanhua. Since then Taiwan tea didn't have to be shipped to Xiamen or Fuzhou for refining. In 1869, Dude sold Taiwan tea directly to the United States.

- 1912 年

中华民国国民政府宣布各省停止向政府进贡茶叶。

1912

The national government of the Republic of China announced that all provinces stopped paying tea as tribute to the government.

- 1915 年

中国多个名茶在巴拿马国际博览会上获奖。

1915

Many famous Chinese teas won awards at Panama International Expo.

- 1926 年

中国丝茶银行在天津正式成立，发行正面有采茶图、反面为缫丝图的纸币。

1926

The China Silk and Tea Industrial Bank was officially established in Tianjin. It issued paper currency with a picture of picking tea on the front and reeling silk on the back.

- 1937 年

中国茶叶股份有限公司在南京成立。

1937

China Tea Co. Ltd. was established in Nanjing.

- 1940 年

复旦大学设立茶学系，这是我国高等院校中的第一个茶叶专科。

1940

The tea science department was established at Fudan University, which is the first tea-focused department in Chinese colleges and universities.

- 1949 年

中国茶叶公司成立。

1949

China Tea Company was established.

- 1958 年

中国农业科学院茶叶研究所成立。

1958

Tea Research Institute, Chinese Academy of Agricultural Sciences was established.

- 1960 年

欧洲茶叶委员会在德国汉堡成立。

1960

European Tea Association was established in Hamburg, Germany.

- 1969 年

联合国粮农组织茶叶协商小组设于罗马。

1969

The FAO Tea Consultation Group was founded in Rome.

- 1991 年

中国茶叶博物馆成立。

1991

China National Tea Museum was established.

- 1992 年

中国茶叶学会和中国茶叶流通协会成立。

1992

China Tea Science Society and China Tea Marketing Association were established.

- 1993 年

中国国际茶文化研究会成立。

1993

China International Tea Culture Institute was established.

- 2015 年

中国茶叶博物馆龙井馆区建成开放。

2015

Longjing branch of China National Tea Museum was completed and opened.

- 2019 年

第74届联合国大会宣布每年5月21日为"国际茶日"。

2019

The 74th UN General Assembly approved in its Resolution (A/RES/74/241) to designate May 21 as International Tea Day.

- 2022 年

"中国传统制茶技艺及其相关习俗"入选联合国教科文组织人类非物质文化遗产代表作名录。

2022

"Traditional tea processing techniques and associated social practices in China" was added to UNESCO's Representative List of the Intangible Cultural Heritage of Humanity.

- 2023 年

"普洱景迈山古茶林文化景观"成功列入《世界遗产名录》。

2023

"The Cultural Landscape of Old Tea Forests of the Jingmai Mountain in Pu'er" was inscribed as a new UNESCO World Heritage.

世界茶业现状
The Global Tea Business Today

全世界有60多个国家和地区产茶，主要集中于北纬49°到南纬33°之间，主产区在亚洲和非洲，其茶叶产量占世界总产量的90%以上。除了中国，目前世界主要产茶国还有印度、肯尼亚、斯里兰卡、孟加拉国、日本、越南、埃塞俄比亚等。

Tea is produced in more than 60 countries and regions around the world, which are mainly between 49° N and 33° S. Asia and Africa are the main production areas, where the tea production accounts for more than 90% of the world's total output of tea. Besides China, other major tea producers are India, Kenya, Sri Lanka, Bangladesh, Japan, Vietnam, Ethiopia, etc.

根据世界茶叶生产的分布，结合气候、生态等条件，可将全世界划分为六大茶区。

Based on production of tea, as well as climate, ecology and other conditions, the world can be divided into six tea-producing areas.

- 东北亚茶区

  Northeast Asian Tea-producing Area

包括中国、日本、韩国等国。

Including China, Japan, South Korea and other countries.

- 南亚茶区

  South Asian Tea-producing Area

包括印度、斯里兰卡、孟加拉国、巴基斯坦等国。

Including India, Sri Lanka, Bangladesh, Pakistan and other countries.

- 东南亚茶区

  Southeast Asian Tea-producing Area

包括印度尼西亚、越南、缅甸、马来西亚等国。

Including Indonesia, Vietnam, Myanmar, Malaysia and other countries.

- 西亚茶区

  West Asian Tea-producing Area

包括格鲁吉亚、阿塞拜疆、土耳其、伊朗等国。

Including Georgia, Azerbaijan, Turkey, Iran and other countries.

- 非洲茶区

  African Tea-producing Area

包括肯尼亚、马拉维、布隆迪、坦桑尼亚、卢旺达、马里等国。

Including Kenya, Malawi, Burundi, Tanzania, Rwanda, Mali and other countries.

- 南美茶区

  South American Tea-producing Area

包括阿根廷、巴西等国。

Including Argentina, Brazil and other countries.

结语
EPILOGUE

茶叶从中国西南的莽莽丛林进入到五彩缤纷的世界，经历了许多考验和波折。

自茶与人类结合开始，茶就以其优良的品质体现出与人类自然亲和的关系，浸润着中华民族的人生理想。

Tea leaves, originating in the dense jungle in the Southwest China, experience many trials, ups and downs before they come into the colorful world.

Since its integration with human, tea has manifested its natural affinity with human with its fine quality, and shone the life ideal of Chinese people.